INDIAN MANUAL OF PLANT ECOLOGY

BY

R. MISRA

M.Sc., Ph.D. (Leeds), F.N.I., F.N.A.Sc., F.B.S.,

Head of Botany Department, University of Saugar.

and

G.S. PURI

M.Sc., Hons. (Punj.), Ph.D., (Luck.), Ph.D. (Lond.),

F.L.S., F.B.S., F.G.S., F.G.M.S.

Forest Ecologist, Forest Research Institute, Dehra Dun.

SCIENTIFIC PUBLISHERS (INDIA)

Print 2024

Published by:

Scientific Publishers (India)
5 A, New Pali Road, P.O. Box 91
Jodhpur 342 001 (India)

E-mail: info@scientificpub.com
Website: www.scientificpub.com

Branch Office
Scientific Publishers (India)
4806/24, Ansari Road, Daryaganj
New Delhi - 110 002 (India)

ISBN: 978-81-7233-795-7

Original Published in India, 1954

Printed in India

FOREWORD

The science of Botany had its beginning in the study of plants in the field. The first phase of this science was devoted to recording the observations of the naturalists and herbalists. The knowledge thus gained was applied to the cultivation of food and medicinal plants.

With the invention of the microscope and the advancement of knowledge in Chemistry, the study of the plant was taken to the laboratory with a view to understanding the structure of the body and functions of the different organs of the plants. Such study has given us a fairly comprehensive knowledge of the plant as an individual organism but such knowledge when applied to the plant in the field has to be modified as various environmental factors have to be taken into account. If the fullest knowledge of the plant is desired, man has to return to the field but armed with the knowledge gathered in the laboratory. The science of Ecology had its origin in such a desire in man.

As soon as we begin the study of the plant in its natural environment, its habitat, we find that knowledge of Botany alone is not enough. Acquaintance with almost all the branches of natural sciences is necessary to study the behaviour of plants in their habitat. Thus, Ecology has to take a comprehensive view, a sort of holistic attitude.

As noted already, the science of Botany began in the field and we have come back to the field after a long digression in search of useful knowledge in the laboratory. Having its origin in man's desire to have fuller knowledge of the plant, the science of Ecology has received an impetus by the recent emphasis put on the dependence of humanity on plants for food and raw materials. Unless we know fully the relation with the environmental factors, exploitation of plants in new surrounding or the fullest utilisation in their familiar surroundings would be fruitless.

Ecology is a new science or rather a new point of view of the Botanical Science. To the uninitiated, there is a jargon of terminology, sometimes conflicting, but definition of concepts is the first essential if co-ordinated progress is to be made in any field of study. Workers in different countries must know clearly what the workers in other parts of the world think and know, otherwise there connot be any progress.

In Ecology, as in other branches of knowledge, the basic principles are international but these principles are to be based on the study of plants in different parts of the world. Any generalisation made from regional observations, however large the region may be, may require modification when applied to another region. Hence the ecological study of a region becomes necessary.

From this stand point, the Indian Manual of Ecology is a welcome publication. The authors are to be complimented for pioneer work. They are conscious of the shortcomings which they promise to remedy in future. In absence of complete information from India, they have had to depend on data collected in other lands, to illustrate the general principles.

This publication is opportune from another point of view adumbrated above, namely, focussing attention on ecological study for the economic exploitation of the plant for food and raw materials for industries. The food crisis in India in recent years has awakened us to the necessity of searching fresh fields and intensifying cultivation with a view to greater production of food, and discovering new plant sources for food supply. Intensive study of plants in their habitats will surely be of help in this direction.

The book should be welcomed by students of botany, agriculturists and foresters alike.

Cuttack. **P. PARIJA**

15th August, 1954

PREFACE

The knowledge of Ecology in India has been mostly confined to structural adaptations of plants in relation to water economy. In most of the universities lectures on ecology are based on teleology, and practical work is relegated to the laboratory wherein the students study anatomy and morphology. In the field they tire out collecting plants and naming them imperfectly. This deplorable position of plant ecology in Indian Universities has been mainly due to two reasons. Firstly, the study was undertaken as a side show and it lacked real leadership in the country. Secondly, the text books available on the subject are by foreign authors, containing foreign observations in which Indian reader does not feel interested for obvious reasons. In order to remedy this situation we have put together our experiences of long teaching and extensive field work in this country.

We are fully aware of the shortcomings of the present work. By devoting a little more time on it and waiting for some more data on Indian vegetation it could have possibly been much improved. However, it is thought more opportune to synchronise its publication with the Symposium on the Vegetation Types of India being held by the Indian Botanical Society. We believe that the gains in focussing the attention of the Indian botanists on plant ecology at an early date shall outweigh the deficiencies of this publication. It is contemplated to bring out the second edition of the book with more Indian material soon.

If the Indian Manual of Plant Ecology succeeds in stimulating the study of causal relations between plants and their environment, the physiological and developmental adaptations of plants and the complex series of vegetational phenomenon on truly scientific basis free from teleology, and create interest in Indian vegetation, our task shall be fully achieved. While putting in much of the data obtained on the Indian soil we admit having freely drawn upon foreign materials also which could not be easily replaced on account of lack of native observations and study.

The present work will be found equally useful to botanists, foresters, agriculturists and every body interested in the conservation of vegetation, soil and water. Probably geographers, soil scientists and meteorologists may also find enough material to interest them. From the academic point of view research workers may find it of special value in locating the gaps in our knowledge of Indian ecology. Teachers may find it worthwhile reorientating the courses of botany in the light of this book.

The plan of the Indian Manual of Plant Ecology in brief, proceeds from a study of autecology to synecology. This is followed by a study of the root and the shoot environments. The general principles of ecology as applied to plants and environments in India have always been kept in the forefront. Nevertheless, sufficient matter of universal application and interest will be found in the text.

No completeness in bibliography of the book is claimed. Neither the references indicate any comparative merit of works done in India. They are just picked up to illustrate running matter of the text. A knowledge of auxiliary sciences for the reader is assumed and the book is not self contained. It is hoped that the teachers will introduce the students into these subjects as and when needed. Further, laboratory methods of plant and soil analysis have been excluded.

It will be appreciated that a work of this character cannot be produced single handed. It is necessarily the outcome of joint efforts on the part of several workers in the field, whose work has been included in the text. We are thankful to them including our teachers, students and colleagues. We are all of obliged to various publishers and authors for their permission to copy some of the figures, tables and photographs. We are especially grateful to Prof. P. Parija for his kind advice so generously given to us, to Shri C. R. Ranganathan, Prof. F. R. Bharucha, Prof. P. Maheshwari, Dr. A. C. Joshi,

Dr. S. N. Das Gupta, Dr. K. A. Chowdhury, Prof. P. N. Mehra, and Prof. Shri Ranjan, for discussing ecological problems with us from time to time which really encouraged us to write this book. The table of contents was circulated for comments to a large number of foresters, soil scientists, agronomists and botanists in India. We are particularly grateful to Shri N.P. Mohan, Shri K.L. Aggarwal, Shri M.B. Raizada, Dr. S. Sinha, Dr. B.M. Johri, Dr. Bahadur Singh, Dr. T.S. Sadasivan, Shri G.S. Lamba, Shri G.S. Dhillon, Prof. Shanti Sarup, Dr S.P. Ray Chaudhuri, Dr. J.K. Basu, Dr. D. Chatterjee, Dr. V. Puri, Shri S.K. Seth for their suggestions.

It remains to add that we owe much inspiration to our common teacher Prof. W.H. Pearsall, F.R.S., to whom we gratefully dedicate this book.

Chakrata
8th June, 1954

R. MISRA
G.S. PURI

CONTENTS

I

Introduction

WHAT IS ECOLOGY

Ecology like economy is derived from the Greek word *Oikos*, meaning house, or a dwelling place; it therefore, deals with the habitat of organisms. The term ecology was employed by Reiter in the same sense; later in 1869, E. Haeckel gave a simple definition of ecology as "the science treating of reciprocal relations of organisms and the external world." In 1895, Warming employed it in the study of plants and defined ecology as the study of organisms in relation to their environment.

The modern definition of ecology is the study of organisms in reciprocal relationship with the environment.

SCOPE OF ECOLOGY

Plant behaviour is regulated by the materials and conditions of the environment and the forces of heredity working within. Since both the environment and the organism are dynamic systems reacting against each other, the resultant organisation has immense potentiality for change. The different factors of the environment may get as much modified by the organisms as they produce effects upon them singly, or collectively. Ecology, indeed, comprehends all such phenomena and envisages control over them.

Depending upon the organisms that are studied ecology is either *Plant Ecology* or *Animal Ecology*. Human Ecology is a branch of animal ecology which merges into sociology, human economics and anthropology. In recent years, interest in human ecology has rapidly increased and the Ecological Society of America held a symposium in 1940 on relations of ecology to human-welfare—the human situation in which a number of interesting observations were made.

When fossil organisms are studied in relation to their past environment the science is called *Paleo-ecology.*

The word ecology is now widely used. In the study of forests, it is designated *forest ecology. Desert ecology, lake ecology, grassland ecology, crop ecology,* etc. are used in the study of different types of organisms and their habitats. In these studies, ecology is emerging into a definite science, with well organised disciplines. It is largely a study of communities and community dynamics and a good deal of knowledge in plant ecology has been gathered.

Some authors, however, consider that ecology is not as definite a science as Botany, Zoology, Morphology, Taxonomy, Physiology, or Genetics, but it is in addition a view-point in the study of organisms. For example, if Morphology and Physiology of plants are studied without any regard to environmental conditions, in which they live and develop, then they form parts of Botany; but if the study of plant structure, or function is conducted in relation to the changing and variable conditions of the environment, then these two divisions of Botany go under ecology. In general practice, both these views are recognised and in addition to being a separate science, ecology is also studied as a "point of view" in many other definite sciences. In this context its field is unlimited, and its activities are varied and many-sided, touching upon several related fields of physical and natural sciences.

Ecology as an organised science is of recent origin, when at the end of the last century geographers began probing into the details of plant distribution and morphologists enquired into structural changes caused by the environment the science of ecology was really born. However, under the influence of Darwinian evolution it soon got overloaded with teleology.

It is only during the last 35 years that the science has been given an experimental and analytical orientation. Now it has assumed a significance that no discipline of biology can ignore.

During the comprehensive stage of ecology it was considered to study merely (1) which species are commonly associated upon similar habitats, (2) what are their physiognomic relations with the environment, (3) why should species be confined to a particular habitat, (4) why do they associate themselves in a particular fashion alone, (5) what type of demands do they make on the environment, and (6) in which way these demands are fulfilled most economically ? But today ecological approach is becoming applicable more and more in a number of fields of human endeavour. Now soil conservation, agronomy, food production, horticulture, flood control, animal husbandry, afforestation, town planning, defence science, human economics, etc., are considered ecological problems. In the 1950 meeting of the British Association for the Advancement of Science, E.J. Russell enumerated a number of problems in which the help of ecology was sought and in his presidential address to the Botany Section of the Indian Science Congress Association in 1942, N. L. Bor had similarly indicated the role that ecology can play in solving India's pressing economic problems.

In ecology is reflected a real unity of science and as such it needs popularisation. It appeals to a large group of workers and provides a common platform for scientists of varied interests.

DIVISIONS OF ECOLOGY

Ecological study of individual species is called *autecology* and that of a population of species growing together *synecology*. Synecology is also called *bio-ecology*, especially when both plants and animals are considered together as a biome. Another term, *plant-sociology*, meaning community relationship of plants, is sometimes used for synecology.

Detailed autecological study based upon extensive culture work led Turesson to formulate *genecology*. It includes a study of structural, developmental and physiological variations of species based upon their genetical potentialities.

Statistical ecology has emerged as a result of detailed analytical and synthetical study of community patterns in grassland and other herbaceous associations of Europe.

BASIC CONCEPTS OF ECOLOGY

The organisms and the environment reacting against each other form a complex unit which should be comprehended as a whole.

Organisms are plastic and in time and space get adapted to a range of environmental variations.

Organisms living together develop an economy of their own and organise into communities.

The environment is a complex of interrelated factors with temporal and spatial variations.

Plant activities such as dispersal, growth, reproduction, death and decay lead to development of vegetaton and change in environmental characteristics. As the changing environment becomes unsuitable for the growth of the preceding communities the composition of the latter gradually changes on account of continued invasions by more aggressive species. Thus, a succession of plant communities at a place is obtained.

Plant succession leads to the development of a climax community which is relatively stable and is in equilibrium with the environment.

STUDY OF ECOLOGY IN INDIA

A knowledge of the flora is a pre-requisite of any ecological study. Hence it was only after the publication of Hooker's flora of India, followed by regional floras, that any attempt of such a study was made.

The forest officers serving in the country were the pioneers in writing out descriptive accounts of the vegetation. However, the causal relations existing between the plants and the environment and the phenomenon of succession were appreciated only after 1920 with the work of Dudgeon (1920). Ecological work on the forests of India was intensified by Troup (1921) on silviculture; Champion (1938), Mohan (1933), and Suri (1933) on forest types and Bor (1936-47), Mooney (1941-50) and Hewetson (1945-54) on regional vegetation. Puri (1948-1954) has recently done a considerable amount of work on the distribution of forest trees in relation to the edaphic factors.

Some work has also been done recently by Sarup and his students on desert ecology (1955).

Hole (1911), Burns, Kulkarni and Godbole (1931), Chakradeo (1922, 1926), Phadnis (1926) and Ranganathan (1938) worked out the ecology of grasslands. Misra and his students (Pandeya, 1952-54, Tiwari, 1953-54, etc.) have recently given data on the phytosociology, soils, succession and autecology of the grasses of Madhya Pradesh.

Bharucha and his students (1940-44) have worked out the biological spectra of the flora of a number of regions. They have also drawn attention to conditions existing in the mangrove vegetation (Navalkar, 1945-48).

The importance of autecological studies has also been realised and studies on some aspects of forest tree species are available (Troup, 1921).

The study of aquatic vegetation, algal ecology and ecology of sand dunes, desert vegetation and miscellaneous habitats have also been pursued in this country. A complete bibliography on the subject is being published by the Indian Botanical Society (Misra and Puri, 1955) and data from the work so far done will be freely used in the pages of this book.

Section A

Plants and Plant Communities

In this section distributional characteristics and growth behaviour of plants (Autecology) and plant communities (Synecology) shall be described.

2

Autecology

Ecological life histories aim at a better understanding of the distribution of species, as well as aid in the evaluation of environment, adaptation and speciation. Much stimulus for such studies has come from the economic importance of certain species and indeed agriculture and silviculture are extensions of autecology. Autecological studies of wild plants have been, however, of greater interest.

The objectives may be varied but in all cases an assessment of potential and the real aggressiveness and susceptibilities of the species under varying conditions of environment are made for this purpose; the means for the regeneration, dispersal, growth and development of plants and their effectiveness are also studied.

The regeneration of a plant depends firstly upon the output of viable seeds, then their percentage germination and vegetative propagation, if any. There are many plants which flower and fruit annually, others do so periodically, and there are still others, which produce seeds only once in their lifetime.

1. Seed output

In table 1 is given the average seed output of a number of species of different habits aud habitats. The data show how plants differ from each other in this respect.

Table 1

Average seed output per plant of given species

Species	Investigator	Av. Seed output	Remarks
1. *Lindenbergia polyantha*	Misra & Rao (1948)	71,731	Annual herb growing on wall.
2. *Anisochilus eriocephalus*	Bakshi (1952)	21,421	Annual herb growing on house tops.
3. *Heylandia latebrosa*	Srivastava (1952)	172	Perennial prostrate herb in open grasslands.
4. *Sida acuta*	Sharma (1953)	8,074	Perennial herb.
5. *Bothriochloa pertusa*	Pandeya (1954)	3,505	Perennial grass.
6. *Dichanthium annulatum*	do	2,795	Perennial grass.
7. *Themeda caudata*	do	1,332	do
8. *Iseilema anthephoroides*	do	463	do
9. *Butea monosperma*	Chourey (1953)	1964	Tree.
10. *Anogeissus latifolia*	,,	19,07,408	,,
11. *Diospyros melanoxylon*	,,	8,746	,,
12. *Lagerstroemia parviflora*	,,	12,926	,,
13. *Tectona grandis*	,,	31,033	,,
14. *Terminalia tomentosa*	,,	8,120	,,
15. *Boswellia serrata*	Sharma (1955)	12,367	,,

It has been observed further that seed output in a plant is dependent upon a number of environmental factors of which light intensity, moisture conditions, biotic influences, diseases, physiological status and age, especially in the case of trees, influence flowering and fruiting. Generally, higher light intensity and dry conditions increase seed output. Pandeya (1954) has shown that the seed output of *Bothriochloa pertusa* and *Dichanthium annulatum* may be considerably reduced in grazed habitats. Sharma (1955) has observed that insects eat away the ovules and young seeds in developing fruits of *Boswellia serrata.* A number of tree species bear more seeds in alternate years or even at longer intervals of time, depending upon the seasonal conditions which might influence the availability of food reserves in the plant, the virus and bacterial diseases and insect populations. Troup (1921) has given a number of cases in which the age of the tree and its growth, either from seeds, or coppice shoots, have influenced the seed bearing capacity. Bamboos flower gregariously at long intervals, and the plants die after flowering as a result of exhaustion of food reserves.

Seed destruction by natural agencies is universal. A large number of animals, big and small, eat plant seeds. Squirrels and rodents are widely active in this respect. Tiwary (1955) has found that a very large percentage of grass seeds are collected and eaten by ants and termites. During dispersal seeds may reach unfavourable situations and may not have a chance to germinate. Seeds of terrestrial plants may get destroyed if they happen to fall in water, and those of aquatic plants may dry upon land.

2. Viability of seeds

When seeds are shed they may remain viable for different periods. In the case of *Shorea robusta* the period of viability does not last longer than a week or so. Seeds of Leguminoseae and many grasses can remain viable for periods, varying from 1 to 15 years.

However, dry and cold storage may keep the seeds alive for longer periods of time. But such conditions are rarely available in nature.

3. Dormancy

Apart from the question of viability the dormancy in seeds and means of breaking it are of importance in autecology. Seeds of a majority of wild plants will not germinate at once after they are shed from the parent plants, even though conditions for germination may be favourable. The causes for this may be varied. Seed coats may be hard and impermeable to water and oxygen; and dormancy in such cases may last as long as the seed coat is not decomposed by natural agencies of water, temperature, bacterial or fungal attacks and become more permeable. Dormancy may be broken artificially by mechanical means or by acid scarification.

In many seeds, dormancy may be due to the presence of a chemical inhibitor of growth. Removal or inactivation of such an inhibitor can terminate dormancy. Washing out of the seed in water may remove the inhibitor in the case of ephemeral weeds. Light and nitrates usually play an important part in the inactivation of the inhibiting factor in germination of seeds.

Dormancy has a survival value for the plant species in as much as it restricts germination to more favourable period for growth. In India, the monsoon is responsible for a shorter wet season, followed by generally a much longer dry season. There are, however, sporadic light showers at long intervals, especially during the cooler dry period from retreating monsoons which provide favourable moisture conditions for germination. If all the seeds were to germinate on such occasions there is every chance for the seedlings to dry up subsequently for want of adequate moisture supply during the unfavourable dry period. However, during the rainy season continued rain for a few months will sustain and establish seedlings. Thus, it is advantageous to have a delayed germination. It is believed that continuous wet conditions such as obtain during the rainy season in India make the seed coat more permeable, or leach out of the seed the chemical inhibitor, thus, helping plants in germination.

In nature seeds get very often buried in the soil. Sarup (1954) has shown that seeds of *Tephrosia purpurea* germinate more easily at 2-4 inches depth than on the surface of the soil. In case of many semi-aquatic plants, like *Ipomea reptans,* and those growing on drying mud, like *Polygonum plebejum, Heliotropium supinum,* etc., the seeds embedded in the mud undergo periodic desiccation and wetting before they can germinate.

The mode of germination and the time taken for the emergence of cotyledons and radicle are specific to plants and has been extensively studied in the case of Indian forest trees.

4. Reproduction capacity of plants

Salisbury (1946) defines reproductive capacity of plants as the product of the average seed output and the fraction represented by the average percentage germination. It is formulated in the following way :—

$$\frac{\text{Average seed output} \times \%\ \text{germination}}{100} = \text{Reproductive capacity.}$$

Reproductive capacity is characteristic of plant species and is indicative of its pressure upon the environment. A high reproductive capacity, usually, ensures a better chance for the survival and dispersal of a plant species. The reproductive capacity of a number of Indian species is given in table 2.

Table 2

The reproductive capacity of some Indian plants

Species	Investigator	Av. seed output	Percentage germination	Reproductive capacity
1. *Lindenbergia polyantha*	Misra & Rao (1948)	71,731	·98	70,266
2. *Anisochilus eriocephalus*	Bakshi (1952)	21,421	6.2	1,328
3. *Heylandia latebrosa*	Srivastava (1952)	17,195	23.4	46
4. *Sida acuta*	Sharma (1953)	8,074	18.5	1,493
5. *Bothrichloa pertusa*	Pandeya (1954)	3,505	43.3	1,517
6. *Dichanthium annulatum*	,,	2,795	45	1,274
7. *Themeda caudata*	,,	1,332	28	372
8. *Iseilema anthephoroides*	,,	463	29	134
9. *Sesbania bispinosa*	Trivedi (1953)			

5. Vegetative propagation

Many plants are capable of rapid vegetative multiplication. Almost all the organs of plants may partake the process. When the undrground parts do so the plant becomes more aggressive and successfully competes with its weaker neighbours. It is because of this fact that the regenerating parts continue to get food supplies of the main stock for a long time, but the plants lack the vigour of sexual reproduction. The successful plants generally have both seed and vegetative reproductions at their disposal. It is true that for wider dispersal seeds are necessary.

Aquatic plants are notorious in shoot regeneration. Bits of *Hydrilla, Potamogeton, Ceratophyllum,* etc. can be easily dispersed to long distances by running water, or birds, and get regenerated rapidly into full plants. Vegetative buds known in temperate regions as winter buds or turions can develop in Indian waters in warm dry periods and multiply the plants with extra-ordinary rapidity. Hence, one should be careful in assessing the reproductive capacity of such plants as seeds may not be so effective in their propagation.

6. Dispersal

Plant species are mobile through their loosened regenerative organs, such as fruits, seeds, bulbs, bulbils, spores and in many aquatic plants bits of shoots. These propagules are carried away from the mother plants by natural agents such as wind, water and animals.

Downy seeds of grasses, such as *Saccharum* and *Erianthus* are carried away by wind to several miles. Flattened pericarp and testa in many fruits and seeds like those of *Ventilago* and several species may reach considerable distances in strong wind. Other light seeds can also be blown up to good distances.

Water current may carry a large assortment of fruits and seeds of land plants, besides the propagules of aquatic plants. Fruits of *Terminalia arjuna* are thus carried to long distances by hill streams which fluctuate in their level. Other fruits and seeds may have aerenchymatous tissues which keep them floating. *Trewia nudiflora* and coconut fruits have aerenchymatous mesocarp and *Carapa moluccensis* has a buoyant testa.

Birds and bats disseminate seeds of *Ficus*, and a large number of species having fleshy fruits. A host of other animals including grazing cattle and goats carry spiny or hooked fruits and seeds attached to their body or by eating fruits and ejecting the seeds along with their faeces and thus disperse them. Almost in all the cases of dissemination the seeds get mechanically or chemically scarified, which facilitates their germination.

7. Seedling growth

The most critical phase in the life cycle of a plant is its seedling stage. During germination and early growth of the embryo the food reserve of the seed is utilised. Hence, heavy and larger seeds with greater amount of stored food have a better chance of producing thrifty seedlings. With the emergence of leaves and the growth of roots the seedling begins to react to a multitude of environmental factors. More often than not, the seedling succumbs to one, or the other of the adverse conditions. Hence a high rate of seedling mortality is observed in nature in a large number of plants.

Any casual observer will find a large number of seedlings drying up in periods of drought. The water supply in the surface layer of the soil falls short of the demand for absorption and transpiration. Only such seedlings that have produced roots long enough to reach deeper and moister levels of the soil have a chance of survival, when the monsoon rains are over in October. Recently, Puri (1954) has shown that a high rate of mortality in seedlings of *Shorea robusta* occurs in Oct.-Dec., the first post-monsoon period.

The shoot and the root of the seedling start living in different environments. The atmospheric conditions may not always be suitable for the life of the young shoot. Diurnal fluctuations in light, temperature, humidity and wind velocity matter a good deal in its growth. The shade cast by neighbouring plants may be sufficient to starve it by cutting down photosynthesis. Nevertheless, many plants grow rapidly with longer internodes and smaller leaves to overtop the neighbours. This phenomenon is more marked in rooted aquatic plants growing in deep water. The low intensity of light in deep water brings about etiolation in *Trapa, Potamogeton, Ceratophyllum,* etc. However, the shoots grow up rapidly to the surface. The complete exposure of the seedling on the other hand may lead to its desiccation by dry winds, even though the soil may be moist. The rapid loss of water sometimes cannot be replaced by slower absorption and conduction. At higher altitudes on the Himalayas, growing buds as a result get desiccated and plants become stunted.

The tender leaves of seedlings are susceptible to a large number of diseases and insect attacks. In the rainy season when it is warm and humid, insect population mounts up. Caterpillars in any number can be seen eating leaves of seedlings of broad leaved species particularly in grassland community. If not already destroyed in this manner the seedlings may be grazed by cattle. Herbivorous animals of all kinds eat up seedlings in large numbers. *Sambhar* and animals of the deer family destroy a large number of tree seedlings, especially of *sal,* adversely affecting its regeneration in *Shorea robusta* foıests of U. P., M. P. and other states.

The seedling has to encounter, perhaps, a more hostile environment at its roots. The demand for water and salts and favourable conditions for their absorption and growth within the soil remain ever so critical. Root competition at different levels of the soil between seedlings and deep-rooted plants for water and salts is always very acute. White ants, termites and other insects attack the young roots most readily. Numerous animals, particularly porcupines eat roots of trees. *Salmalia malabarica* is the worst sufferer in this respect and many matchwood plantations have been ruined in this way.

Misra (1938) has shown that the roots of the rooted aquatic plants are as important for absorption of salts from the mud as they are in the case of land plants. Hence in any programe of autecological study of rooted aquatic plants attention has necessarily to be paid to their roots.

8. Vegetative growth

With the establishment of the seedling, further growth follows depending upon the intensity of the environmental factors in any localitity. Any critical factor in the soil, or the atmosphere leads to a certain degree of adaptation in the plant. Some of these adaptations are structural in character, such as variations in the size of leaves, stems, tissues, cells, etc. Others are physiological in nature, such as variations in osmotic concentration of the cell sap, respiration, photosynthesis, growth., and still others are connected with phenology and habit of the plants. Species differ in the degree of plasticity with regard to exhibiting one or the other character. However, the extremes of environmental factors limit the distribution of a species by exhausting its capacity for adaptation. In autecology, therefore, we try to know the range of tolerance of a species.

9. Reproductive growth

For a successful growth in time and space, species must flower and fruit. Vegetative reproduction is decidedly not so efficient to perpetuate the species. It is only the sexual reproduction which gives chance for intermingling of genes and producing variations and mutations—thus extending the scope of adaptation.

The environment has a marked effect upon the flowering and fruiting of plants. The initiation of flowering is dependent upon the length of the day and the temperature to which the growing plant is exposed. In relation to the length of the day, plants may be classified into the following three groups :—

(i) Short day plants flower only when the daily period of illumination is shorter than a critical photoperiod of illumination. *Nicotiana tabacum* (var. maryland mammoth) is an example.

(ii) Long day plants flower when the period of illumination exceeds the critical duration. *Spinacea oleraceae* is a long day plant.

(iii) Day neutral plants can flower in a wide range of photoperiods. *Zea mays* and *Capsicum annum* are examples.

A sufficient amount of data for Indian species and wild plants are lacking with regard to their photoperiodic behaviour, but it is easy to see that long day plants will mostly be confined to higher latitudes, where day lengths are longer during the growing summer period. The short day plants on the other hand may flourish nearer equator and the day neutrals may be found anywhere.

Similarly, a number of plants in colder regions flower because of their exposure to low temperature for some time, during the year. Photoperiodic and vernalisation data have recently been published by Chronica Botanica.

10. Autecological studies

A number of interesting features of such studies will now be given. Misra and Rao (1948) have shown that *Lindenbergia polyantha* is an ecotype of *L. urtaecifolia.* The former is calcicolous—growing on lime rich soils and most abundantly on old cracked walls. By culture work, it was shown that the calcium content of the plant and that of the soil are both directly correlated. According to Bakshi (1952) *Anisochelus eriocephalus* grows on the tiled roof of houses as the loose soil between the tiles is better aerated and does not get waterlogged during the rainy season. However, under similar conditions the plant is not found on the ground on account of interspecific competetion. *Heylandia latebrosa* according to Srivastava (1952) can grow in grazed grasslands where it often escapes injury on account of its prostrate habit. It has a high requirement of light intensity. Sharma (1952) observed *Sida acuta* growing on lime rich soil though it is not a calcicole, because it escapes competition from other plants, which become aggressive to it on lime poor soils. Thus, a large number of interesting facts about plant behaviour can be revealed from autecological studies.

11. Genecology

Extensive culture work with a number of species collected from different parts of their distributional range showed Turesson (1922) morphological and physiological variations, which were very often intergrading and not found distinct enough to warrant their separation into discrete species. Thus, he recognised a number of races with morphological and physiological characteristics within a species—further, he found that many of the genetic races are correlated with habitats and are true ecological races. Based upon these data, he recognised the following types within the Linnaean species :—

1. **Ecophenes,** are the plants which show variations in size, and number of leaves, shoots, flowers, etc., and in habit, but belong to the same genetic stock. For example, *Euphorbia hirta* growing in shade and moist places is an ecophene of the plant growing in exposed drier places. Salisbury used the term *ecad* for *ecophene.*

2. **Ecotypes** or ecologic races are genetically distinct, but on account of their interfertility they are often put together under the same taxonomic unit. *Lindenbergia polyantha* was supposed to be a starved form of *L. urtaecifolia* by Cooke (1908), but Misra and Rao (1948) have shown that they are distinct ecotypes. The first named species is a calcicole and the other grows on ordinary soil. When growing together they interbreed and produce intergrading forms, which taxonomically merge the species into each other. There seem to be a large number of ecological races among Indian plants that await detailed investigation.

   Turesson's eco-species and coeno-species are higher entities of his classificatory units of ecotypes.

12. Plant indicators

It is clear that plant species by their limited distribution indicate certain conditions of the environment. To that extent each species can be used as an indicator of its habitat. Thus,

Urochloa reptans and *Sporobolus* indicate overgrazed grassland, *Saccharum spontaneum* indicates poor soil drainage, *Capparis spinosa* and *Carissa spinarum* are indicators of intense soil erosion in forest areas, etc., etc. *Pteridium* indicates burnt and highly disturbed conifer forests in the Western Himalayas, and *Phoenix acaulis* indicates similar conditions in *Shorea robusta* forests of Madhya Pradesh *Strobilanthes* and *Impatiens* indicate too moist conditions for silver fir regeneration in conifer forests. Hole (1911-1913) worked out a number of grasses in the forests of Uttar Pradesh that are indicators of conditions for successful and unsuccessful regeneration of *Shorea robusta*. Plant indicator concept in ecology is of great value in forest reproduction work in India and deserves a special study.

The ecological life history of a species is studied in relation to variations in the environmental factors. Field observations are very often verified by culture experiments. Pelton (1951) has outlined schemes of autecological studies for herbaceous and tree species. However, the details can be varied in each species according to the objectives. The British Ecological Society has recently sponsored an extensive scheme of autecological work on the flora of the British Isles. In India, a good deal of work has been published on forest trees.

Plate I - Nelumbium-Nymphaea-Eleocharis community, Sagar lake.

G.S.P., April, 1952

To face p. 20.

Plate Ia - A grassland community at sagar

R.M., October, 1953.

To face p 21.

3

Synecology

1. Nature of the community

Plants grow in association with each other. A population of similar or dissimilar species growing together and mutually interacting upon one another constitute a plant community. In a community, plants share the same habitat. Their interdependence may assume different forms. For example, tall plants in a community may provide shade to plants of lower life forms and there may develop in some cases a competition for light. Temperature and humidity may be altered in a community by different growth; competitions for absorption of water and salt from the soil is a common feature in a community. A competition for shoot and root space may also arise in a closed or dense community. The dependence of a parasite, such as dodder or *Loranthus* upon the host is an extreme case of competition for survival. Epiphytism and climbing habit in plants offer a lesser dependence in a plant community.

Symbiosis is of common occurrence in many communities —one way benefit of supplying fixed nitrogen to the soil is offered by leguminous plants to non-legumes. The decomposition of forest litter in the same manner may benefit others by ameliorating soil conditions in a community.

An interesting example of the dependence of one plant upon another may be provided by the growth of *Striga* on grass roots. The grass roots produce chemical stimulants, which helps in germination of *Striga* seeds close to them. The seedling of the parasite then get attached to the host.

There may be chemical inhibitors produced by certain species retarding the growth of others in a community.

In an open communiy, plants grow singly and quite apart from one another ; and the interdependence of the plants may be less obvious in such cases. Nevertheless, on account of the uniformity of the environment the community is recognised and here too under suitable conditions a closed community may develop.

One of the familiar examples of a community is a stand of forest trees with an undergrowth. Savannah with tall grasses and scattered trees, grassy meadows with a number of small flowering plants, rough grazing pastures, roadside and footpath vegetation, and a group of plants in a pond or lake are all plant communities of different types and characters (Plate I).

There may be smaller communities within a larger community such as a ground flora community or a community of ferns, liverworts and mosses growing on a stone or tree bark in a forest community of trees (Plate II).

The term community in the abstract is, therefore, applicable to any group of plants whether they are large or small and belong to one life form or the other, but are essentially growing in an uniform environment.

A community possesses structure, development, history and behaviour. All these community characteristics are correlated with environmental factors just as individual plants and species are correlated.

2. Structure of plant community

The structure of community is dependent upon the physiognomy of the individuals composing it and their disposition towards each other. A complex mature community like that of a forest (Plate III) has a few overtopping species which by their bulk and growth modify the habitat and control the growth of other species. These are called the dominants.

Plate I-A mixed moist deciduous community at Konsao, Dehra Dun.

G. S.P., December, 1953.

To face p 22.

Plate II-A ground floor community of Bracken - *Mercurialis* in *Quercus robur* community in Whippandale woods

G. S.P., August, 1947 To face p 23.

In a community there may be one over-topping species or individual dominant or more. Those which share supremacy are termed co-dominants. In addition there are a number of secondary species in a community.

Besides, composition and dominance, the community possesses a number of layers or storeys related to the height groups of plants. Thus, there may be the topmost layer or a layer or two of smaller trees in the middle; a layer of shrubs and small trees ; a layer of tall herbs, tree seedlings, short herbs, prostrate plants and mosses, etc. This stratification is absent in a single layered stand of similar plants such as in a crop field. In a uniform regular forest attempt is made through management techniques to produce single storeyed plant communities.

The different species composing a community may have different patterns of distribution or dispersion. Some individuals may be aggregated together and others may grow apart from one another. Although, no two bits of the same community are identical in this respect, ~~however,~~ some degree of repetition of similar stands in a community may be discernible.

A community is called open when the individuals composing it occur apart from one another and the ground is bare at places for the invasion of new species (Plate IV). When the growth becomes dense, with a little or no space between plants, and the entry of invaders in the community is reduced to the minimum, it is termed as a closed community.

3. Origin and development of a plant community

A plant community with its particular environment constitutes an entity which has its origin and development. Since the area around the community is open to the growth and invasion of plants for a very long time it may seem difficult to appreciate their pressure upon it. The processes concerned with the origin of plant community can, however, be better appreciated by the study of a piece of land, newly exposed by landslide, erosion, deposition or made barren by a catastrophic agency.

The seeds and propagules of the surrounding vegetation reach such areas all the time but in the changed set up of the environment only some of them germinate and grow up quickly. As is usually the case, many of these seedlings and plants perish under the stress of the environment, but those few, which are capable of growing in exposed dry situations, survive, reproduce and multiply.

The process of successful growth of plants that have migrated from the neighbouring areas to new situations is called *ecesis*. Plant migrations are the universal phenomena, leading to the *colonisation* of new areas.

In the process of rapid colonisation, individuals of pioneer plants are usually aggregated, which inevitably leads to interspecific and intra-specific competetion for space and food, during different phases of their growth. By these actions and interactions on the habitat the early colonists modify their own environment and the equilibrium is upset.

New plants enter gaps and may succeed eventually in ousting out some of the early colonists and preparing ground for their own exit, subsequently, for the invasion of other colonists, better suited to the new habitat. In this community all functions of the growth and development are visible. The change proceeds indefinitely into an orderly succession, when series of communities enter the habitat, linger on and then quit, making room for others, till a stage is reached when one community comes to stay and both the habitat and the community have reached a more or less stabilisation.

It must be understood that the processes of migration, ecesis, colonisation, aggregation, competetion, invasion, reaction, interaction and stabilisation, as described above, are continuous phenomena, taking place everywhere and in all communities. They become more evident in barren areas and open habitats than in closed habitats. It must also be understood that there is no strict sequence of these processes as different plants may be subject to different forces at the same time.

Plate IIa. - A community of mosses in the English Lake District.

C.S.P. To face 24.

4. Succession

Even the brief description of the plant community given above will indicate its dynamic character. The community—environment system is ever unstable, seeking new and progressive equilibria of relative stability.

Succession in Clement's terminology is the natural process by which the same locality becomes successively colonised by different groups, or communities of plants. This occupation of the area from the beginning to the end is marked by a series, or procession of communities, and on the whole, the process is continuous. Sometimes, when a mature plant community passes from one into the other, the change is clear and well marked.

However, a time comes when the processsion of communities merges into a stable, or climax community, which tends to remain in harmony with the environmental complex for a long time. The ephemeral communities leading to the development of the climax are called seral communities, which togeher form a 'sere'. This process of succession is indicated in fig. 1. *(See page 26)*

Nature of succession. Succession is studied in two major types of habitats, which correspond to the sere and the climax. Sere is generally controlled by edaphic relations, while climax is climate-controlled. The seral habitats are small and more numerous ; the climax on the other hand are large and very few, limited to only one habitat in a region of uniform climatic conditions. The main difference between these two habitats is that sere is an open habitat, while climax is more or less closed. In seral habitat there is always invasion and colonisation, while little or no fresh invasion occurs in a climax habitat.

Seral stages are like the waves, which occur and recur at various intervals, but when the climax comes, it comes to stay. Both habitats are dynamic and largely progressive ; sere being edaphic is mostly progressive, unless there is a biotic disturbance.

Climax is reached only after a very long time and is generally in terms of geological ages. Intermediate successional

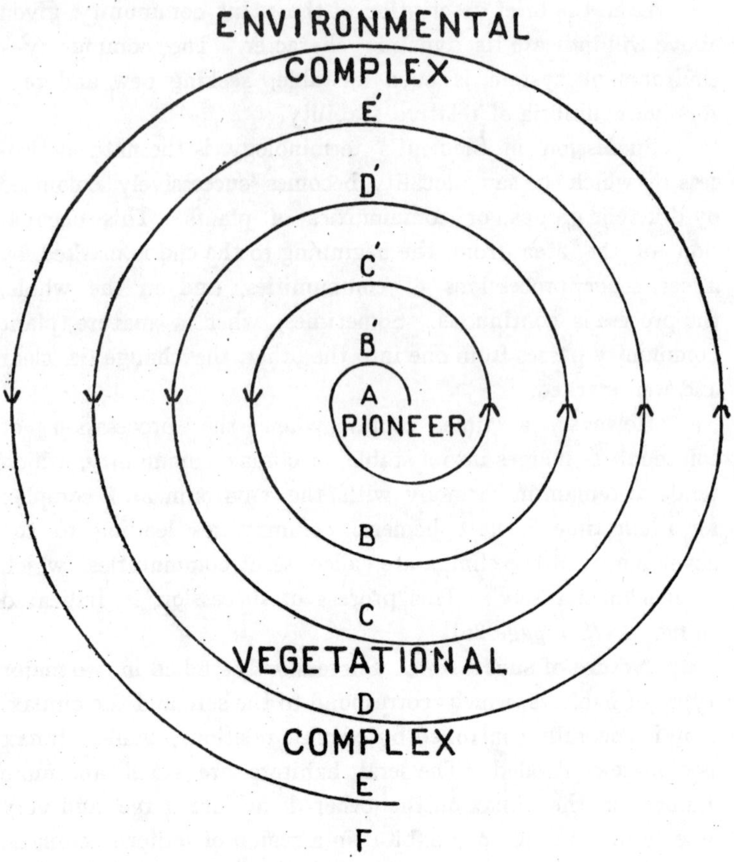

Fig. 1: SUCCESSION DIAGRAM

A–E seral communities; F. climax community.

stages, which may be arrested in their development for long periods of time, may appear to be climax, but actually they are not. The word climax is unfortunate, since it gives a false idea of static conditions, and goes against the very nature of ecological concept, which is essentially dynamic. The most appropriate word for the climax community should be final or relatively stable community at the present time. On account

complex forest community of shorea robusta at Asarori, with dominant species in the upper story, a well developed shrab layer and regeneration of different tree species.

G. S.P., December, 1952

To face p 26

←Plate II

Plate IV-An open community of *Acacia calechu* in boulder bed in the river Song at Lachiwala, Dehra Dun. Note open spaces between trees.

G. S.P.

To face p 27.

of the interactions of man, world's climax vegetation has been largely destroyed and what we get now are bio-edaphic communities arrested in their development by the human influences.

Process of succession. The first essential requisite in the process of succession on any bare area is the migration of plants from the surrounding areas and their aggregation together on the scene of activity. As the bare space is almost always larger than the numbers of new arrivals, there is no congestion to begin with, but each one of the migrants reacts on the soil and when it has taken hold of it, tends to gain equilibrium with the climatic conditions, immediately surrounding it. The pioneer plants draw freely upon light of the atmosphere and when they have sufficiently grown, they restrict the coming in of later plants by casting shade, occupying more space, or by taking in available supplies of water, etc.

In the early stages of succession, the pioneers show cooperation and after having adjusted themselves to the environment they begin to multiply. This is the beginning of the competition among plants of the same species, and in different species, for nutrition from the atmosphere and the soil. In this process of struggle the number of plants that survive and occupy the ground is decreased. There is further decrease in the number of these pioneers as they react on the environment and change it. In most cases, this changed environment becomes unsuitable for the growth and spread of the existing plants, but opens up avenues for the coming in of fresh arrivals.

In terrestrial habitats, pioneers and usually low organisms which can stand scarcity of water, need less soil nitrogen and are thrifty in nature. C/N in the tissues of these plants is usually high. These are light demanders, and are usually exacting for soil minerals. Their seeds and fruits are wind dispersed, or are usually small and attract insects and birds for dispersion to other areas.

By their death and decay, the pioneers increase organic matter, moisture and nitrogen content of the soil, which now becomes suitable for the growth of the next lot of invaders. In this way development of the community progresses and the number of new arrivals at each stage goes on decreasing till the climax stage is reached, when the existing plants attain, such an equilibrium with the environment that the entry of any new arrival is extremely limited.

Causes of succession. Ecology concerns itself primarily with causes and processes. In the dynamic universe some cause initiates a change in the environment, which produces a corresponding effect in organisms. The organisms then react on the environment, or in turn become a secondary cause which produces a secondary effect. In this same way the effect becomes cause again and produces still another effect in turn, and so on. This phenomenon of cause and effect, as already referred to above, is rhythmical and cyclic in nature and marks changes which are universal. The dynamic ecology studies this cycle of cause and effect in plants and in their environment.

The chief cause of succession in a habitat is climatic. A bare area is produced by a change in the climate or sometimes by biotic agencies, then the initiation of a sere begins. Large scale landslide, flooding by river, destruction of vegetation by fire, etc. are some of the apparent causes of the starting of succession in vegetation. Secondary succession is set up by partial destruction of the vegetation by any one, or all of the above mentioned agencies.

A succession after its initiation or partial development may be deflected from its normal course by such disturbing factors, as fire, grazing, or other biotic agencies, and may be called a *deflected succession.*

For development of the vegetation in an area, both climate and soil shape the nature of the plant community, its structure and form. In seral habitat, however, succession from one community to the other goes on becoming increasingly dependent on soil conditions, till a stage is reached, when the plant community attain equilibrium with both factors of the environment, the climate and the soil. The vegetation then enters the climax stage.

Kinds of succession. Successions may be of several types, depending upon the type of habitat on which they start, and are manifested differently in different seres. We may recognise the following types of seres :

(i) Hydrosere. This is a sere originating in water. The pioneers in a lake or pond are phytoplanktons. As they gradually die, decompose, and mix up with the silt brought from the surrounding land by rain water and arising from the shore due to wave action, a soft mud is formed at the bottom. This now forms a suitable substratum for the growth of rooted plants and *Potamogeton* spp., *Ceratophyllum, Hydrilla, Vallisneria,* etc. may grow in shallow regions, where light can easily penetrate the water. These plants contribute to further building up of the substratum which is colonised by plants with floating leaves like *Nymphaea, Nelumbium, Limnanthemum, Aponogeton, Monocharia, Trapa,* etc. The water by this time acquires a higher concentration of salts and organic matter and is thus fit for the growth of free floating plants, like *Pistia, Azolla, Lemna, Spirodella, Wolffia,* etc. This rapid growth, especially near the margin, is soon accompanied by a number of sedges and reeds converting the area into a swamp or marsh. All the vegetable matter and silt will raise up the bottom gradually from the centre to the margin of the lake. As the mud is exposed to the atmosphere its aeration brings about conversion of reduced nutrients like ammonia, suphides, etc, to nitrates, sulphates, etc. As a result, there is invasion of the neighbouring terrestrial plants and the aquatic communities are replaced step by step by the climax community characteristic of the area.

(ii) Lithosere. This sere initiates on rock or detached blocks and stones. The initial plants are characterised in most cases by mosses and lichens, which occur on dry mineral substrata, devoid of any moisture or organic matter. These plants are able to fix atmospheric nitrogen and moisture and by their death and decay produce soil that conserves organic matter, moisture, and nitrates. On this substratum, there develops a variety of small herbs and by the growth and death of these low forms, some soil accumulates on the rock. Due to the action of organic acids produced by the decomposition of plant parts and atmospheric changes, the rock on the surface decomposes, producing a hospitable habitat for more exacting species of shrubs and trees, to colonise and develop in this sere in later stages. The invading species in almost all stages of their colonisation and early development are dependent upon the mineral composition of the rock and the climatic conditions, which make the minerals in an available form.

Although, this sere develops on rocks or hard surfaces, side by side in crevices of the rocks and other well developed parts of the habitat, in addition to lowly organised plants, representatives of higher forms are also present in the initial stages. Thus, tree seedlings, chiefly with mycorrhizal roots, and xerophytic types are invariably found with low forms of plants in early stages of invasion. The succession discussed in the lithosere develops rather quickly in the latter habitats. The chief action of plant growth in the lithosere is the increase of the depth of the soil, the amount of moisture and organic matter content and nitrogen in the surface layers of the soil.

(iii) Psammosere. This starts on sand. The pioneers are generally sand binding grasses with runners,such as *Spinifex* and other plants like *Ipomea biloba.* After sand has been stabilised, a variety of shrubs and trees colonise the area. The fixation of sand and increase in organic content and moisture retaining capacity of the soil are the main trends of the habitat in this sere.

(iv) Halosere. This is a sere beginning in salt water or on saltish soil. The initial stages in this sere are characterised by small plants, as in Psammosere with the difference that the plants in this case are able to withstand large amounts of salt in the soil.

(v) Xerosere. This is initiated on bare areas, whether they are rocks, sands or stones, but have extreme deficiency of moisture in the soil.

(vi) Serule. This is a miniature succession of microscopic organioms, such as bacteria and diflerent types of fungi on fallen logs, twigs or tree litter.

(vii) Eosere. The development of vegetation in an *Eon* or *Era* is called *Eosere.* It is a major developmental series within a climatic climax of the geological era. The term *Geosere* is sometimes used to designate the total plant succession of the geological past.

In all the above mentioned cases, succession starts from primitive substrata, devoid of any plant propagules. Hence, these seres are classed as 'Prisere' or primary succession. But when succession starts from partially built up substrata with some pre-existing plants or plant parts it is a case of 'sub-sere' or secondary succession.

In most of the cases, it is the vegetation, which continues to modify the habit, during succession. Hence, its course is known as 'autogenic'. However, under intense erosion or depositions extraneous forces seem to be more powerful than the pioneer plants and in such cases the topographic course of succession is 'allogenic'.

Some ecologists have talked of "retrogressive succession" in a vegetation, which is degenerating on account of continued biotic exploitation. They have visualised the existence of a succession of communities of lower denomination. However, it should be clearly understood that at every stage of destruction secondary developmental successions are set up in the community and what is happening is merely stoppage of the real secondary successions, and further degeneration of the community. Succession is thus always developmental and cannot be visualised, otherwise.

Dudgeon (1920) thought of 'seasonal succession' in the meadow lands of Uttar Pradesh. But this expression like the retrogressive succession is equally misleading, Seasonal aspects are recurrent and not developmental in the same sense as real successions is. Hence, Misra (1946) has regarded the perennial species forming the basic community, and the annuals representing only aspect communities, in the meadow vegetation of such areas.

CLIMAX

The idea of climax in vegetational development was first conceived by Hult in 1885, but it was left to later investigators, like Cowles, Oliver and Clements to give it a concrete shape.

In 1916, Clements presented it as a complex organism, inseparably connected with its climate. So, climax came to be known as major unit of vegetation under the prevailing conditions of climate. In this sense, *climax* is always used with reference to climate and is called *climatic climax.*

The climax vegetation is also called *formation.* A plant formation or climax community is a unit of vegetation in which main relations of the dominants with the climate are identical. It is a product of climate and is strictly delimited and controlled by it. In India, we may recognise 4 main formations, as tropical, subtropical, temperate and alpine.

When both animals and plants are considered in a formation, the climax is usually called as *biome.* Sometimes, the word climax is used for a plant community, which is a smaller subdivision of a formation.

Nature of the climax. The climax has the following characterstics:

(i) **Unity.** Climax is not only a response to a particular climate but it is an indicator and expression of it; so just as a knowledge of climate will indicate a climax, a climax community will likewise give indication of its climatic conditions. The dominant plant forms in a climax signify unity of the climax and these are the best expressions of the climate. For example, in a grassland climax, grasses are dominant, in a scrub, and shrubs and in a high forest trees are the dominant life-forms. Trees as dominat life-forms may occur in a number of climaxes, as tropical, temperate, etc.

The life-from in a climax is reflected in its species, so the unity of the climax is represented by its species which sometimes signify relationship with the climate. The reactions and coactions of the dominants in a climax determine the nature of the secondary and subordinate species. The secondary and subordinate species may change in a climax locally, in response to local differences in soil, biotic influences, or by the micro-climate.

(ii) Stabilization and change. A climax attains stability with the environment in a long time, sometimes in thousands and millions of years. In areas where the development of vegetation is continually disturbed, or disrupted by natural or artificial causes, under the same climate, a real and stable climax may not reach at all.

The dominant species of the climax for all intent and purposes are fairly stable and in their growth do not modify the habitat to any large extent, so as to bring about a change in the unity of the climax. However, these species exhibit seasonal or periodic changes, which do not leave any permanent impress on the habitat. Seasonal or periodic change in the dominant species determine the nature of the ground flora forms including tree seedlings and growth of dominant and secondary species. In any climate, climax does not change unless the climate changes.

However, climaxes can be destroyed by human activity and in a very greatly disturbed area it may seem that two or more climax formations are existing in apparent equilibrium with one and the same climate. This is, however, rare.

(iii) Origin and relationship of the climax. Each climax, in addition to having its own distinctive growth and development characteristics, is derived from some other preceding climax. Thus, a climax has both ontogeny and phylogeny, which can be studied in the same area. By the study of fossil remains it is possible to know the origin of climaxes. For example, the study of fossil plants in India has shown that temperate climax is in existence in the Kashmir Himalayas, since the Pleistocene and tropical climax is existing in the Peninsular India since the Tertiary period.

Cooper believes that the development of vegetation takes place like the development of a stream, which coming through mountainous country divides, unites, bifurcates and is visible only in parts. It is not possible for anyone standing at one place to get a complete view of the stream and the hidden parts may look quite mysterious. The origin of vegetation climaxes like the stream is in the remote past, in time and space.

Some ecologists regard that the development of vegetation takes place in a series of climaxes, while others, conversant with dynamic geology, consider it to have occurred in a steady evolutionary way, interrupted frequently by sudden revolutionary geological changes. The latter view has probably more evidence.

(iv) Test of a climax. Major climaxes as tropical, Temperate, etc., are distinct units and clearly recognisable but there are others, which are not so well marked. The extent of the climax is tested by the extent of its certain dominant species, which must be present throughout, in all its subdivisions. These species must either be always dominant in all the subdivisions or they may be dominant in only some. Subdominants may also be common in the various divisions of the climax and may link the parts of the climax occurring at different places. The acid test of a climax is that under the existing climate it must maintain its chief physiognomic characters and must always reproduce itself indefinitely.

If we accept Clements' definition of climax, being inseparably connected with the climate, then under the existing climatic conditions there could be only one climax and all other forms of vegetation, which look like climax should be subordinate to it. Various terms, such as pro-climax, dis-climax, pre-climax, sub-climax, post-climax and ser-climax have been proposed to differentiate these communities from the climax.

Pro-climax is more or less a stable community resembling climax formation in some respects, but it is capable of undergoing a physiognomic change by disturbances. If the modifying influence is chiefly the activities of man or animals then such a modified climax is called *dis-climax,* such as a grassland in a forest area. Very often a strip of vegetation of a higher life form is found within a climax on account of the presence of more mesophytic situations, such as a forest along a stream in a grassland climate. The higher type in such cases is known as *post-climax.*

Pre-climax or sub-climax is a climax of lower life forms from the one adjacent to it, and results from different edaphic conditions within the climax.

If in an early stage of development a climax is arrested in its development and is held stationary for an indefinite period such a climax is called *ser-climax.*

The concept of climax is extremely complex and as our knowledge of vegetational change is never complete there has been a good deal of difference of opinion in interpreting the climax concept. There are a number of ecologists, distinguished in their own way, who believe that under one climate in an area there can be more than one climax, which is opposed to Clements' original idea of the climax. They recognise an edaphic climax, due to soil conditions, a biotic climax, due to human influences, and a physiographic climax, due to physiography; thus laying the foundation of a *poly-climax* theory in vegetational development.

In fact, biotic and edaphic climaxes are not in any way different from Clements' dis-climax and pre-climax communities. As long as the real nature of succession and climax is understood it matters little whether one believes in *mono-climax or poly-climax* for the sake of convenience in field.

Some recent work in India and abroad leads one to believe that a biotic climax, edaphic climax, pre-climax, or sub-climax are nothing more than successional or seral stages, which are indefinitely arrested in their development by geological, soil, or biotic conditions (Puri, 1950). These local patches in the landscape do not seriously alter the general picture. Huxley has suggested the name of "embryonic communities" for them.

CLASSIFICATION OF PLANT COMMUNITIES

Plant communities have been classified in various ways by ecologists, depending upon their view-points, and emphasis on one character, or the other.

Warming (1909) based his classification on water relations and divided communities into hydrophytes, mesophytes and xerophytes. Schimper (1898) recognised forests, grasslands and deserts, according to climatic zones. Clements (1916) having a clear grasp of the development of communities evolved a complete system of classification. According to him vegetation occurs in climax as well as in developmental or seral units. To distinguish between the two types of units, Clements has given different names. For climax units, he proposes the suffixation and for seral communities suffixies is used. The climax unit is subdivided into associations and the developmental unit is divided into associes. The definition of various terms used in the two units are given below :-

Association is a climax unit, with two or more dominant forms. Associations are those climax communities, which are associated in a region to form a formation with a uniform taxonomic composition. Different associations in a formation may have at places slightly different floristic composition, but all these must agree with the formation in physiognomy and development.

Associes is a developmental equivalent of the association and being a seral or successional stage is replaced by other plant communities in succession.

Consociation is a climax community with a single dominant. This is of the rank of an association with the difference that it has a single dominant, which in some cases may have a low distribution and exercises little control over the climax community, as a whole.

Consocies is a developmental community with only one dominant species. In a consociation, consocies is controlled by facies.

Faciation is a subdivision of association, which is related within a general climate to small differences in moisture relations and temperature.

Facies is a developmental community in the sense of faciation.

Lociation is a localised variant of an association differing from it in the composition of some main subdominant and chief secondary species.

Locies is a developmental community in the sense of lociation.

Society is a community characterised by one or more subdominants. These are almost always species of different life forms from the dominants. Society is mostly used for ground flora communities.

Socies is a seral community of the rank of society.

Clan is a group of plants of secondary species in local or restricted small areas. It is a permanent feature and has a distinctive character, being frequently the result of vegetative propagation.

The Clementsian system of classification so elaborate and complete however, could not be followed with success everywhere. In America with many virgin forests and a short history of human activities, it was easy of application. But on the Continent vegetation has been so much modified by man that ecologists found it difficult to stick to a number of terms, which could be applied to hypothetical or potential communities alone.

The Montpellier-Zurich school of ecologists led by Pavillard and Braun Blanquet evolved floristic method of classification. It is based upon the belief that every species has a greater or lesser indicator value. However, since the social value of the species differs, floristic lists cannot be compared mechanically. Hence they have taken mainly characteristic species into account. The most closely related associations are united into "alliance". Alliances are further grouped into order and orders into "class." Classes form "circle of vegetation," which is the highest unit of the system.

The one redeeming feature in these complex systems of classification is that all ecologists agree on the term 'association' as the unit of vegetation, although, the association of most authors differs according to their view point.

As it has been already stated, the association of Clements is a climax community. Flahault and Schroters' (1910) definition that "an association is a plant community of definite floristic composition" is still widely accepted on the continent. Gleason (1917) presented the individualistic concept of association. According to him the association is a temporary and uctuating phen omenon depending upon the selective action of the environment and the surrounding vegetation. Cain (1947) has very much supported this view and Pandeya (1954) has successfully employed the unit in classifying the grassland vegetation in India.

A reference may be made here to Gams' (1918) unit the 'Synusium' which has been used by Saxton (1922) to describe the seasonal vegetation in India. Synusia refer to similar life forms and habitat and each layer in a community has been regarded a synusium. However, very few ecologists have found any use for this unit.

4

Methods in the study of Plant Communities

The pioneers in ecological research used mainly floristic methods, which consisted in listing the number of species and genera present in a particular habitat. The vegetation was characterised by the general features of the characteristic flora. The floristic methods were applied in large geographical areas and a geographical, or regional approach to the study of vegetation was given by such an eminent botanist as Schimper.

In India, a large number of floristic studies were conducted and finally Hooker (1906) divided the country into a number of geographic regions. This system was recently modified slightly by Calder (1937) and Chatterjee (1939). Chatterjee in making ten broad divisions of Indian vegetation took into consideration the endemic features of the vegetation of this subcontinent. His divisions are N. W. Himalayas, Central Himalayas, E. Himalayas, Deccan, Malabar, Indus Plain, Gangetic Plain, Rajputana desert, Assam and Burma.

Although, the floristic methods provided a broad classification of the flora and vegetation they lacked precise information on plant communities, their composition, structure, growth, etc. The study of community structure and its composition was largely developed in Europe. The Zurich-Montpellier school of vegetational analysis headed by J. Braun—Blanquet has developed detailed methods and it is largely due to the efforts of this school that there emerged in Europe a phase of ecology, known as phyto-sociology. Braun Blanquet's book on the subject gives a detailed analysis of vegetation by phyto-sociological methods. The main basis of the study by these methods is a quadrat. A *quadrat* is usually a square, circular, or rectangular area for a detailed analysis of vegetation.

Phyto-sociological methods, which originated and developed in Europe, are now greatly in use throughout the world. They are used in comparing one plant community with another; for showing changes in community structure or composition from time to time ; and in the study of succession. The uses of these methods are almost unlimited, having been employed both in fundamental and applied ecology.

Analytical Methods

The ecological analysis of plant communities is conducted by running one or a number of transects at right angles to fertility gradient lines in an area to be studied. This method is used throughout the world, not only in forest ecology, but also in agricultural ecology, soil survey and enumeration studies in forestry. The chief merit of the transect method is its simplicity. It is statistically planned and provides quantitative data, which are used to draw qualitative inferences regarding dynamics of vegetation, succession, etc.

A *transect* is a line, or a belt, of a suitable width, passing through an area to be studied, across the main topographical, geological and geomorphological features. Basically, it is not much different from forester's strip, which is laid across the stand in volume enumeration surveys. It can be compared to Finney's "Sampling Unit". The transect may either be laid as a line, or a belt, passing through a plant community, or it may be rectangular unit of any width, from a few feet to a couple of chains, or more. The length of a transect is usually not fixed in advance, but depending upon the nature of the study, the structure, composition and type of vegetation, and physiographic features of the area a transect

Plate IV - Shola forest and grassland on horizontally startified rock in the Nilgiris at Ootacamund.

G. S.P., January, 1950. To face p 40.

may be from a few furlongs to a few miles long. Puri (1952-53) has described in detail the transect method and the following is the summary of his papers.

The most important thing to take care in laying a transect is to fix its proper direction. This is fixed after a thorough reconnaisance of the area and with due regard to fertility gradients. As fertility lines in a forest are usually related to contour lines, which mostly coincide with changes in environmental complex, in hilly areas the transect must be laid at right angles to contours. In the Himalayas, variations in climatic conditions, chiefly temperature, relative humidity, rainfall and in some cases soil climate follow changes in altitude. Therefore, a transect from lower to higher altitude, or *vice versa*, will not only sample vegetation types but will also show the relationship of vegetation with the environmental conditions.

In hills formed of metamorphosed rocks, or unstratified deposits of homogeneous composition the changes in climate are very much related to changes in level. On plateaux, which are horizontally stratified, as are found in some parts of the Central India, changes in climatic conditions are not so well marked with changes in level. Both sides seem to have similar environmental conditions as is shown in Fig. 2. In the Deccan

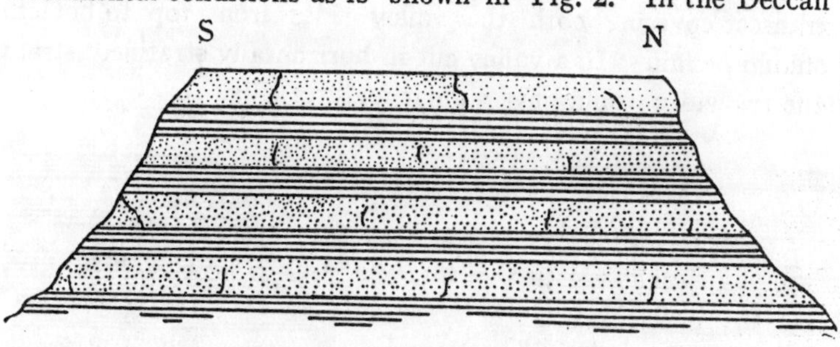

Fig. 2:—A Plateau Structure.

plateau which is slightly tilted towards the east, changes in the climate with the rise of land are again not so distinct in small homogeneous areas. As one proceeds from the equator towards higher latitudes changes in climate with regard to altitude become more marked. In all these areas, where geological features are not complex, a transect should be laid in north-south, or in some cases in N.E.-S.W. direction.

In hills formed of stratified rocks, the strata of which dip towards one side, or the other, forming two distinct—scarp and dip-slopes, all contour lines do not necessarily coincide with fertility lines. In such cases a transect must only be run parallel to the dip of the strata, covering both the dip and the scarp slopes (Fig. 3). In a strike valley the two sides will bear

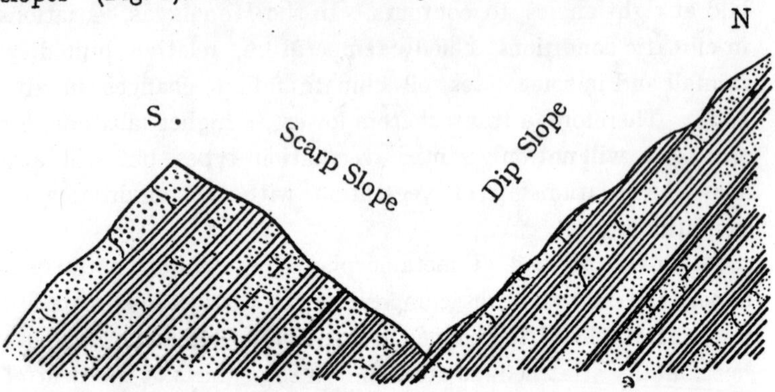

Fig 3:—A valley in the Himalayas showin g Dip and Scarp slopes

different] types of vegetation, for the study of which a transect covering both the valley sides from top to bottom should be laid. In a valley cut in horizontally stratified strata the two sides would have similar features (Fig. 4).

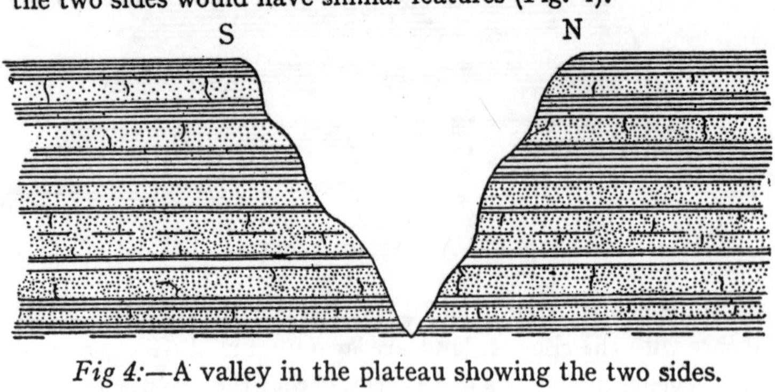

Fig 4:—A valley in the plateau showing the two sides.

In still another case, physiography of the area may be

complicated by the presence of secondary deposits, such as, alluvia, flood-plain deposits, moraines, volcanic lava, mud-flows, dykes, etc., which may be present without any regard to contour lines. Fertility gradients in such cases are extremely complicated and a number of small separate transects running N.-S. may have to be studied, if differences brought about by the heterogeneous features of the environment are to be properly appreciated.

In hilly areas, a transect must cover at least one topographic unit, which consists of three valleys, three tops, three north slopes, and three south slopes (Fig. 5). In gently sloping

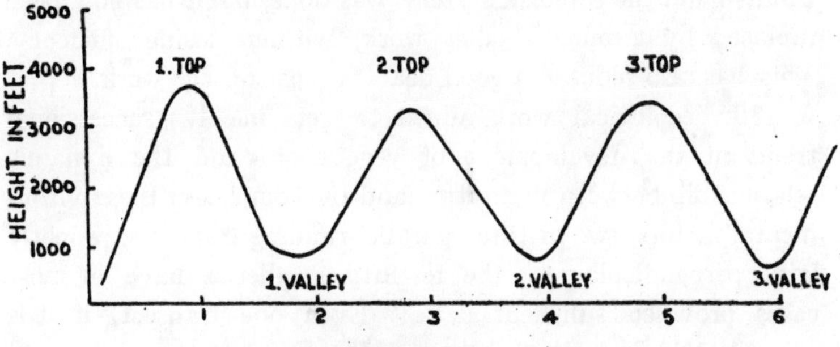

Fig. 5:—A topographic unit in ecological study.

areas or those that are dead level the length of the transect may be a mile, or more, so that a sufficient amount of variations in the vegetation are sampled.

In forest crops managed under regular or uniform systems of management the length of the transect should be longer than in a primeval forest. A successional transect should again be longer than the one which is designed to study only the progress of regeneration in a small area. In the study of ground flora communities and tree seedling-growth the transect is usually much smaller in length.

In grassland surveys, or in the study of weed flora in an agricultural crop the transect may be in the form of a number of small disconnected rectangular or circular plots in a line.

In soil survey, or soil conservation and land management studies a transect has to be continuous and long, covering a number of soil series and groups.

In national forest surveys of Finland, Sweden and other north European countries continuous transects were laid from north to south and east to west at a distance of three to four chains. Thus, the total enumeration of vegetation and detailed charting of land forms were done.

In enumeration surveys, a number of parallel running transects are usually studied and between 10-25% enumeration is done. In early days, a cent per cent enumeration like the Finnish and the Swedish systems was done, but it has now been replaced by a much smaller work, without losing efficiency. This has also reduced a good deal of cost of the work.

In ecological work aimed to get merely successional trend in the development of vegetation and the dynamic relationship between vegetation and the complex of the environmental factors, two or three parallel-running transects, properly laid, perpendicular to the fertility gradients, have in most cases provided sufficient data. Even one transect, if it is properly laid and planned, is considered by some workers to have yielded sufficient data for ecological purposes. Thus, Morrison and his co-workers (1948) were able to bring representative data by the study of a single transect in Africa.

In volume enumeration or other surveys, in which accurate quantitative estimate of vegetation is desired, randomised method of selecting area for running transects is used and a similar method is employed in agricultural survey for estimating yield of crops or in grassland studies. But for forest successional studies selection of sites for laying a transect is usually done in one or two different strata. In some cases, this transect gives better results than randomised sampling ; in most cases it gives an accurate data. Only in a very few cases the data obtained by selection method may not provide sufficiently accurate results, but such cases have to be studied and compared.

The actual recording of vegetation and the study of vegetational environment is done in small units along a transect. These units are called quadrats which in Finny's terminology are "Recording Units". The word quadrat means a square, but in actual practice this may either be a square, a rectangle, or a circle of a convenient size.

The size of a quadrat depends upon the type of vegetation that one is dealing with, and the size of the field party. For recording trees and shrubs circular plots of 18.6 feet radius are convenient in thick forests for a small party of four or five men. In open forest with little shrub layer and sparse ground flora a party of five to six men may be able to chart conveniently a quadrat as large as with a radius of 37.2 feet. In respect to acreage the quadrat with 18.6 feet radius is approximately 1/40 of an acre, and in the other with a radius of 37.2 feet an area equal to approximately 1/10 of an acre is sampled. In double or three storeyed forests, as are met with in tropical regions, where visibility in a line is small, and where the ground layer is profusely developed, the radius of the quadrats may even be halved, or further reduced. Ground flora communities and tree seedlings are recorded in smaller quadrats, usually of one meter square area. These are located in the larger tree quadrats. In grassland communities, or in the study of weeds and agricultural crops the size of a quadrat is nearly 1/3 or 1/4 of that used for ground flora species in a forest. In the study of smaller forms of vegetation, for example, mosses, liverworts lichens, etc. the size of a quadrat is much reduced and may not be more than one foot square.

While for general ecological studies the size and number of quadrats to be examined in a community may not be of very great importance, for comparing two different plant communities it is necessary that the size of a quadrat and the number of quadrats to be studied should be fixed for different types of vegetation. The aim in such detailed work is to record all, or nearly all, the species in the area by as few quadrats as feasible. As this has not been studied for various types of forest in this country information has to be collected at an early date. Some data on these lines are given in the accompanying diagrams. Fig. 6 shows the relationship of

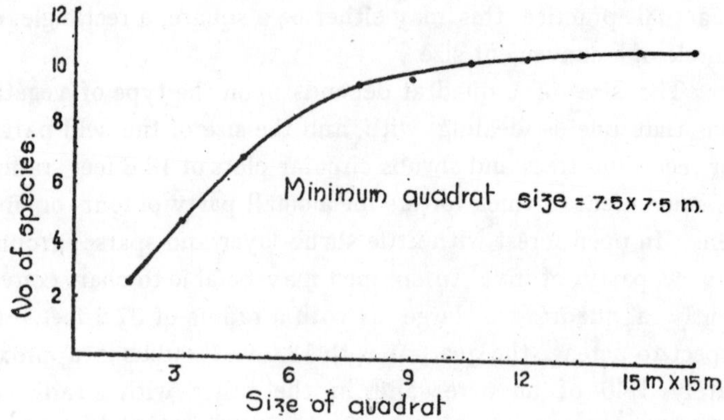

Fig. 6. Species area curve for Patharia forest (Bhatia, 1954)

size of quadrats with the number of species. The minimum size of the quadrat to get the maximum number of species is indicated. In Fig. 7 number of 7·5×7·5 meter quadrats and

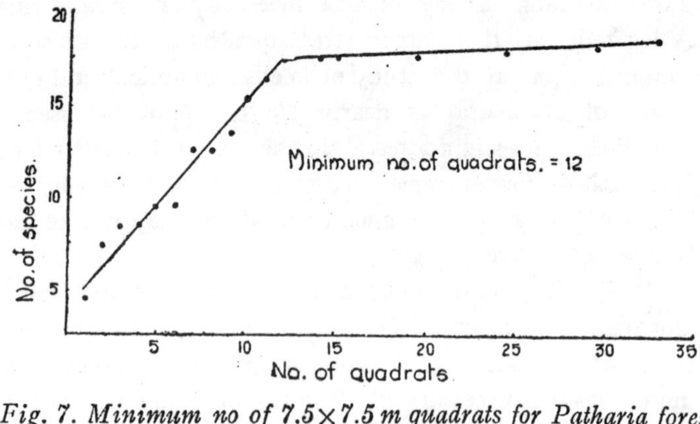

Fig. 7. Minimum no of 7.5×7.5 m quadrats for Patharia forest. (Bhatia, 1954)

number of species occuring in these are plotted and the minimum number of quadrats that should be studied is indicated.

The quadrats are studied in a transect at suitable intervals, care being taken that two adjoining quadrats do not overlap and the same plants are not counted twice. The distance between two quadrats again depends upon the aim of the study.

The transect method outlined here may be familiar to foresters by different names. For example, Lowdermilk (1927) suggested miliacre areas of 6.6 feet square for counting regeneration. Kadambi's (1943) spot method consists of circular areas or spots of one chain radius for complete enumeration.

Depending upon the study in view, quadrats may be temporary or permanent. Permanent quadrats are marked by pegs or fenced by wire, or poles for repeated observations and record. Quadrats may be of the following types :—

(i) **List count quadrat** is most commonly used for listing species and counting their numbers. Such quadrats are useful for studying the floristic composition of a community.

(ii) **Chart quadrat** is used for recording detailed and to-scale growth and distribution of species. All the plants are faithfully recorded at intervals of some months, or years and changes in the structure, pattern and composition of a plant community are studied. The data obtained at intervals of a certain period provide successional trends in vegetational development. This method is also useful for comparing two apparently similar communities. As the charting is to be done in detail, accurate mechanical methods are usually employed for such studies. A pantograph is often used for this work. For more accurate and faithful representation paired pictures taken at intervals of a fixed time are used.

(iii) **Experimental quadrat** is mostly permanent and is employed to record the growth, survival, invasion and spread of plants composing a community. Aspect or periodicity of vegetation is usually studied in an experimental quadrat. In table 3 and 4 *(see pages 51 and 52)* charting of two quadrats in the Lachiwala Forests is shown. In most cases experimental quadrat is designed to study, in addition to vegetation, the environmental factors. In such quadrats, thermometer, rain gauge, atmometer, thermo-and hygrographs are installed to study the community climate. The interactions of various biotic factors in a community are also studied in experimental quadrats. Permanent quadrats are marked in a forest for the study of (a) autecology of species, (b) progress of the regeneration, (c) seasonal variations in microclimate, (d) seasonal and with-depth variations in soil climate, (e) progress of decomposition of tree litter and to mark soil fertility trends, (f) ground microflora and fauna, (g) competition, co-ordination, and dependence

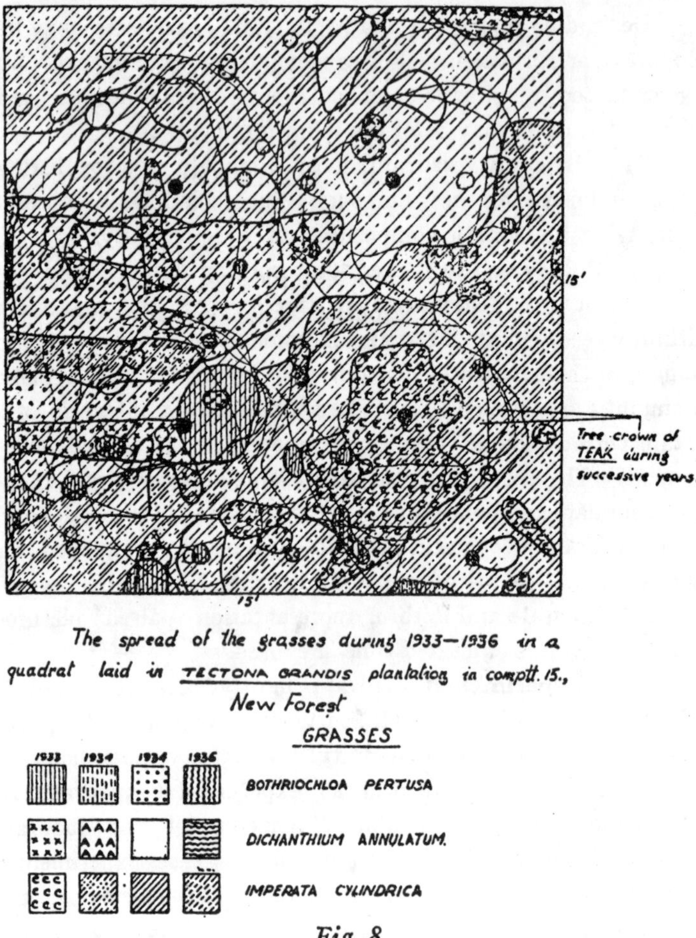

The spread of the grasses during 1933—1936 in a quadrat laid in TECTONA GRANDIS plantation in comptt. 15., New Forest

Fig. 8

of various organisms in different strata of a community and (h) growth and development of various plants and, so on.

Records of two permanent experimental quadrats from New Forest, Dehra Dun, showing the changes in the spread of ground flora species with the spread of tree crown are shown in Figs. 8 and 9.

The spread of the four ground flora species during 1928-1932 in a quadrat laid in SHOREA ROBUSTA plantation in comptt. 19., New Forest.

Fig. 9

Sometimes a *grid pattern quadrat* is employed for a long term study of vegetation in adjoining situations. A grid pattern quadrat is shown in Fig. 10.

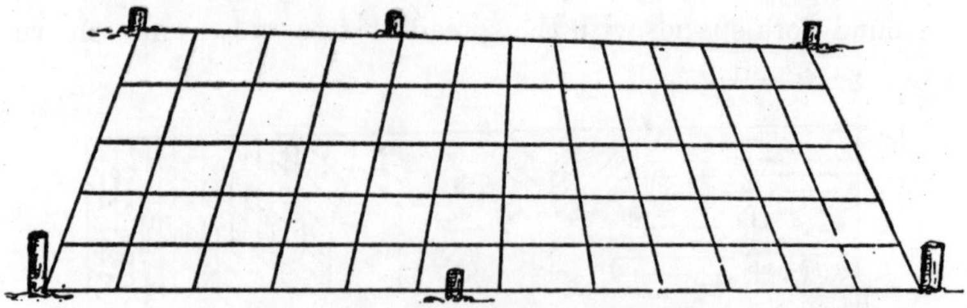

Fig. 10:—Grid pattern quadrat.

The most interesting and useful part of the transect study is the thoroughness with which the recording of vegetation and evaluation of vegetation environment are done in various types of quadrats. Along a transect, quadrats are generally temporary and in a tree quadrat all the trees of different species and in various strata or layers are counted separately and are recorded in separate columns of a suitable form. The design of a form with recorded data from two types of forests are shown on pages 51 and 52. The diameter or girth of trees at breast height is measured, the number of poles and saplings of each species is separately recorded. Felled, lopped, burnt, or malformed trees are specially mentioned to indicate human interference that the forest has been subjected to.

Phenological conditions of the dominant species in the community are also recorded. If any tree is diseased or infected with fungi, or insects, a special mention of this is made. In addition, a sample of such diseased specimen is collected for detailed examination.

In addition to the quantitative data, qualitative observations on cover, position of tree crowns in relation to light intensity, sociability, or gregariousness of the species, their stratification, their vitality, etc. are made.

In a tree quadrat one or more small one meter ground flora quadrats are studied. Usually, the position of

TABLE 3

A chart of quadrats in the Lachiwala forests.

Date *5-4-1955*.

Forest name or number.	... Sal Comptt. 7a.	Geological horizon.	... Middle Siwaliks.	Seepage.	... Towards N.E.
Virgin/Secondary.	... Secondary.	Secondary deposit/ Rock *in situ*.	... Siwalik clay and conglomerate *in situ*.	Aspect.	... Northern generally, but changing.
Fellings. (Dates).	... October 1954.	Dip of the rock.	... N. E.	Rainfall conditions.	... About 90 in annually.
Fire.	... Had occurred.	Slope.	... Fairly steep, generally 1 in 3 or 4 feet.	Temperature conditions	.. Spring.
Grazing.	... Permitted.			Soil moisture conditions	.. Dry.
System of management.	... Uniform.			Soil type.	... Clayey, generally with conglomerate outcrops.
Phenological conditions at the time of study	.... Leaf fall in *Shorea*.				

Transect No. 1 from N.W. to S.E. Quadrats of 15 feet radius at intervals of 40 ft.

Quadrat No.		1	2	3	4	5	6	7	8	9	10
Flants :—											
I. Upper Storey :—											
Shorea robusta	...	5	5	22	-	38	46	-	60	55	26
Terminalia belerica	...	5	-	-	4	-	-	-	-	-	-
Terminalia tomentosa	...	-	-	2	-	8	8	-	-	5	-
Lagerstroemia parviflora	...	-	-	4	13	-	-	-	-	-	-
Adina cordifolia	...	-	-	-	-	4	-	-	-	2	-
Anogeissus latifolia	...	-	-	-	2	-	-	-	5	-	-
Buchanania latifolia	...	-	-	-	-	3	-	-	3	-	-
II. Middle Storey :—											
Mallotus philippinensis	...	44	24	16	-	-	-	5	25s	40	63
Ougeinia dalbergioides	...	5	-	-	22	8	-	5	5	-	-
Semecarpus anacardium	...	-	-	-	-	-	8s	-	-	-	-
Grewia vestita	...	-	-	-	4	-	-	-	-	-	-
Lannea grandis	..	-	-	-	4	-	-	-	-	-	-
Phyllanthus emblica	...	-	-	-	-	3	-	-	1	-	-
Cassia fistula	...	-	-	1	-	-	-	1	-	-	-
III. Shrub layer and climbers :—											
Colebrookia oppositifolia	...	+	+	+	+	-	+	-	-	+	+
Holarrhena antidysenterica	...	-	-	27s	-	8	-	-	-	-	-
Zizyphus xylopyra	...	-	-	2	-	-	-	-	-	-	-
Randia dumetorum	...	-	-	-	-	8	-	-	-	-	-
Bauhinia vahlii	...	-	-	-	2	-	-	-	1	-	-
Millettia auriculata	...	-	-	+	-	+	+	+	-	-	+
Clerodendron infortunatum	...	+	+	+	-	+	-	+	+	-	+
IV. Ground flora and tree seedlings :—											
Oplismenus sp.	...	+	+	-	+	-	+	+	+	-	-
Adiantum sp.	...	-	+	+	+	-	-	+	+	+	+
Shorea robusta seedlings	...	+	+	+	-	+	+	-	+	+	+
Terminalia seedlings	...	-	-	+	-	+	-	-	-	+	-

* S=Saplings

Quadrat No.	Girth, diameter, height, etc.; vitality, dispersion, etc., of the dominant species.
1	*Shorea robusta* girth–1′-5″, 1′-1″, 1′-0″, 2′-0″, 1′-7″.
2	*Shorea robusta* girth–1′-1″, 1′-4″, 8″, 2′-4″, 1′-4″.
3	*Shorea robusta* girth–5″, 8″, 6″, 1′-1″, 1′-3″, 7″, 1′-7″, 1′-2″, 11″, 1′-3″.
4	
5	*Shorea robusta* girth–1′-3″, 1′-0″, 6″, 1′-1″,5″,1′-4″, 8″, 1′-2″, 1′-1″, 11″.
6	*Shorea robusta* girth–6″, 1′-1″, 1′-2″, 1′-5″. 10″, 7″, 11″, 1′-3″, 1′-0″, 9″.
7	
8	*Shorea robusta* girth–1′-4″,1′-1″,1′-3″,1′-7″,1′-2″,11″, 7″ 1′-6″, 1-1″, 6″, 9″.
9	*Shorea robusta* girth–7″, 1′-1″, 1′-0″, 8″, 1′-5″, 1′-2″, 11″, 8″,1′-1″,1′-3″, 6″.
10	*Shorea robusta* girth–1′-1″,5″,1′-0″,1′-3″, 6″,10″,1′-3″,1′-6″, 1′-7″,11″,9″,1′11″.

Quadrat No.	Topography, geology, soil, humus conditions, etc.
1	On a slope down the ridge, soil is fairly moist and clayey, leaf litter fairly decomposed.
2	Further down the slope, soil here is less moist and clayey, leaf litter less decomposed.
3	On a rising slope, soil less moist and clayey, leaf litter of sal and decomposing.
4	Further up the slope, soil fairly moist and clayey loam, leaf litter fairly decomposed.
5	On the sloping ridge, soil less moist and clayey, litter mainly of sal and decomposing.
6	On a sloping ridge, soil less moist and clayey, leaf litter decomposing.
7	Down the slope, soil less moist and clayey, leaf litter mainly of sal.
8	Further down the slope, soil less moist and clayey, leaf litter less decomposed.
9	On a rising slope, soil less moist, clayey and bouldery, leaf litter less decomposed.
10	On a plain ridge, soil fairly moist and clayey loam, leaf litter fairly decomposed.

Quadrat No.	Notes on the forest biota, etc.
1	Lopping and grazing fairly common.
2	Lopping and grazing fairly common.
3	Felling has been done, grazing fairly common.
4	Felling previously done, termite mound present.
5	Felling, lopping, grazing seem to be common.
6	Felling, grazing, etc., seem to be fairly common.
7	Termite mound present, grazing fairly common.
8	Grazing and lopping seem to be fairly common.
9	Lopping and grazing fairly common.
10	Felling and grazing fairly common, termite mound present.

TABLE 4

A chart of quadrats in the forest of the Bashahar Himalayas.

Date *5–7–1955.*

Forest name or number.	... Undemarcated fir forest.	Locality.	... Daran Ghati, Upper Bashahar Div.	Seepage.	... Only on the dip slope.
Virgin/Secondary.	... Secondary.	Geological formation.	... Quartzites and granites.	Aspect.	... S. W.
Fellings. (dates).	... Not done.	Secondary deposit	Nil.	Rainfall conditions.	... About 80" annually.
Fire	... Less common.	Rock *in situ.*	... Quartzites and Schists.	Temperature conditions	.. Monsoonic.
Grazing.	... Common.	Dip of the rock.	... N. E.	Soil moisture conditions	... Moist.
System of management.	... Selection.	Slope.	...Fairly steep with 1 in 5 gradient.	Soil type.	... Immature soil.
Phenological conditions at the time of study.	Evergreen trees, mainly. Deciduous spp. in leaves.				

Transect No. 4 from S. W. to N. E. Quadrats of 10 feet radius at intervals of 25 ft.

Quadrat No.		1	2	3	4	5	6	7	8	9	10
Plants :—											
I. Upper Storey :—											
Abies webbiana	...	7	7	4	3	2	2	1	3	2	1
Picea smithiana	...	-	-	-	1	1	2	-	-	2	-
Pinus excelsa	...	-	-	-	-	-	1	-	-	-	-
II. Middle Storey :—											
Quercus semecarpifolia	...	2	1	-	-	14	-	3	1	12	-
Acer villosum	...	3	1	-	-	-	1	-	-	-	-
Acer cultratum	...	1	-	-	-	-	-	-	-	-	-
Corylus sp.	...	-	-	1	-	-	-	-	-	-	-
Prunus padus	...	-	1	-	-	-	-	-	1	-	1
Rhododendron arboreum	...	-	-	-	1	-	-	-	-	-	-
Juglans regia	...	-	-	-	1	-	-	-	-	-	-
Salix sp.	...	-	-	-	-	-	-	1	-	-	-
Taxus baccata	...	-	3	3	-	-	6	-	1	-	2
Ilex dipyrena	...	-	-	-	1	-	-	-	-	-	-
III. Shrub layer and climbers :—											
Viburnum cotinifolium	...	-	-	-	-	-	1	-	-	-	1
Viburnum stellulatum	...	-	-	-	-	-	-	-	1	-	-
Jasminum humile	...	-	-	-	-	-	+	-	-	-	+
Cotoneaster bacillaris	...	-	-	-	1	-	-	-	1	-	-
Vitis sp.	...	-	-	+	-	-	-	-	+	+	-
Rosa sp.	...	-	-	-	-	-	+	-	-	-	-
Smilax vaginata	...	-	-	-	-	+	-	-	-	+	+
IV. Ground flora and tree seedlings :—											
Senecio sp.	...	+	+	+	+	+	+	+	+	+	+
Viola sp.	...	+	+	-	-	-	-	+	-	-	-
Geranium wallichii	...	-	+	+	-	-	-	+	-	+	-
Ainsliaea aptera	...	-	-	-	+	-	-	-	-	-	-
Fragaria vesca	..	-	-	-	-	-	-	-	+	-	+
Impatiens sp.	...	-	-	-	-	+	+	-	+	-	-
Rumex sp.	...	-	-	-	-	-	-	+	-	+	-
Valeriana sp.	...	-	-	-	+	+	+	-	+	-	
Quercus seedlings	...	-	+	+	-	-	-	-	-	+	-
Rhododendron seedling	...	-	-	-	-	-	-	+	-	-	+
Lastrea sp.	...	+	-	-	-	+	+	-	+	+	-

Quadrat No.	Girth, diameter, height, etc.; vitality, dispersion, etc., of the dominant tree species.	Topography, geology, soil, humus conditions, etc.	Notes on the forest biota, etc.
1	,, girth–4'-3",	Further down the slope, soil moist and clayey, leaf litter composed of broad leaves and needles decomposing.	Lopping and grazing fairly common.
2	,, girth–3'-7'', 3'-1''.	Down the slope, soil moist and clayey with shale pieces. Leaf litter much decomposed.	Grazing fairly common.
3	,, girth–3'-3", 2'-9", 2'-7".	Further down the slope, soil moist and clayey with shales interbedded. Leaf litter decomposing.	Grazing and lopping common.
4	,, girth–2'-5".	Down the slope, soil clayey and moist, leaf litter much decomposed.	Grazing fairly common.
5	,, girth–3'-11", 2'-0".	On the slope down the road, soil clayey and moist, leaf litter much decomposed.	Lopping and grazing seems to be fairly common.
6	,, girth–1'-5", 3'-10".	Down the slope, soil clayey and moist, leaf litter composed of broad leaves and needle decomposing.	Grazing, and lopping seems to be fairly common.
7	,, girth–2'-0'', 2'-0'', 1'-7''.	Further down the slope, soil sandy clay and fairly moist, leaf litter fairly decomposed.	Grazing seems to be fairly common.
8	,, girth–4'-8'', 1'-4'', 2'-0'', 2'-3''.	Down the slope, soil sandy clay and fairly moist, leaf litter much decomposed..	Lopping and grazing rare.
9	,, girth–1'-9'', 2'-3'', 2'-9'', 1'-5'', 1'-11'', 2'-1'', 1'-7''.	Further down the slope, soil sandy clay and fairly moist leaf litter fairly decomposed.	Grazing fairly common, lopping rare.
10	*Abies webbiana* girth–2'-3'', 2'-0'', 1'-4'', 1'-5'', 2'-7'',2'-9'', 1'-11''.	On the slope above the road, soil sandy clay and moist, leaf litter fairly decomposed.	Grazing seems to be common, lopping is also seen here and there.

these quadrats is fixed by randomised technique but some bias is always made in their selection. These quadrats are studied to evaluate the composition of ground flora communities and their relationship with the tree communities on the one hand and tree-seedling-growth on the other. On account of this, ground flora quadrats are carefully selected so as to contain in one meter area a pure, or an almost pure patch of a particular ground flora community and its associated tree seedlings. Sometimes when a sufficiently large patch of a single ground flora community is not present then a mixed community is taken into the quadrat. The effort is always made to study in these quadrats tree seedling-growth so that vitality of the dominant species and the trend of succession could be properly assessed.

The ground flora community is named after its dominant one or two species which are usually *abundant* and are recorded by the symbol (a), the secondary species are recorded by such symbols as *common* (c), *rare* (r), *present* (+) and *absent* (—). In regeneration counts only established seedlings are mentioned, though recruitment is invariably indicated.

In addition to ground flora community and tree seedling-growth, data on tree litter, humus, soil colour, texture, visible micro-fauna and micro-flora (worms, termites, insects, fungus growth, etc.) are collected.

The data on forest environment are collected by recording temperature of the tree community at various heights and of the ground flora communities at the soil and plant levels. Relative humidity is recorded and wind velocity and other features of the climate are determined. The soil climate is also determined by taking temperature of the surface and subsoil with earth thermometers. Samples of humus from the top 0"-6" of the soil, or from the rooted regions of the tree seedling and ground flora community are collected from the centre of the ground flora quadrat.

To study the conditions of the subsoil and development in relation to tree vegetation in the quadrat, one or more soil profiles are examined. In soils with well developed horizons morphological characters, including composition, colour, texture, etc. of A and B (or F and H) layers and presence of roots, are carefully examined. In azonal soils, the whole profile at various depths is examined for morphological characters. In such soils, samples are collected from the depth of 0", 6", 12", 24", 36", and in some cases from 60" and 72". These samples are carefully packed in tins or paper bags, and transported to the laboratory for detailed mechanical and chemical analysis. For microbiological analysis the soil samples from the upper layer are collected in air-tight containers so that characters of the soil do not get changed during transit by excessive oxidation. Some simple soil tests for the measurements of moisture, pH, base status and nitrate production are carried in field.

General geographical features of the area are examined and if there is any marked change, it is carefully noted.

In the Himalayas and in some parts of Madhya Pradesh, Bombay and Deccan, where geological conditions are heterogeneous a change in geological formation may occur rapidly, sometimes within a distance of a few feet. As the changes in geological features may mark pronounced changes in vegetation special care is necessary to observe minute variations in the rock and soil in each quadrat along the transect.

Depending upon the type of work in view and the nature of the forest, quadrats are studied at intervals of 40-50 feet.

A number of quadrats are recorded in the way described above and the data on vegetation are analysed in the following way :—

The number of all the trees, saplings, shrubs, etc. occuring in a tree quadrat are added up and percentage of each species is determined. This provides an idea of the density of the different species in a quadrat and when density figures for the different species in all quadrats in the area are worked out density classes are separated.

Frequency of a species is determined by finding out the percentage of the total number of quadrats in which it occurs. Frequency occurences are determined for all the species in a transect separately, or in all the transects studied in the area together. These are grouped in Raunkiaer's (1934) five frequency classes as follows :—

Frequency class A= 1 - 20%
B=21 - 40%
C=41 - 60%
D=61 - 80%
E=81 - 100%

On the basis of 8,000 frequency percentages determined by Raunkiaer he found that in an area frequency class A included 53% of the species; B 14%; C 9%; D 8%; and E 16%. This is given in Raunkiaer's normal frequency diagram. In Fig. 11 is shown frequency diagram of the forest at Patharia on Raunkiaer's lines. In Fig. 12 is given frequency diagram of a grassland community.

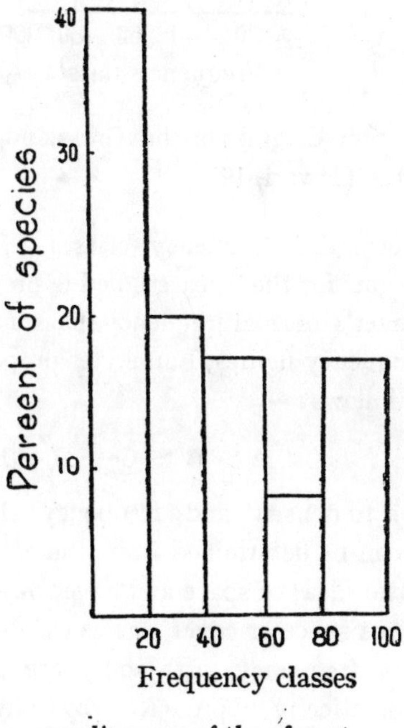

Fig. 11 :—Frequency diagram of the forest complex, Patharia Hill, Sagar. (Misra & Joshi, 1952)

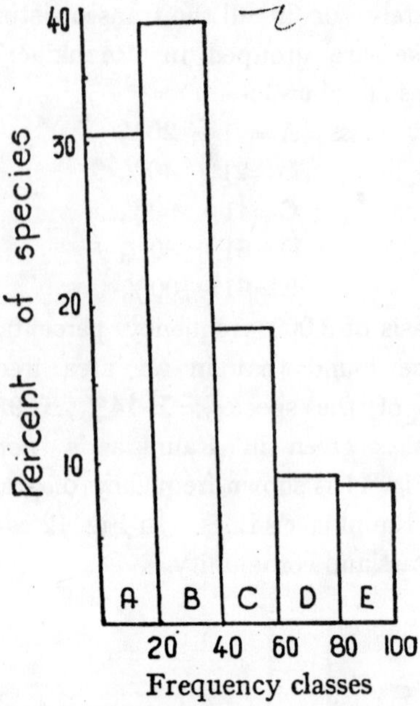

Fig. 12 :—Frequency diagram of the Grassland community of Sagar. (Tiwari, 1952)

After determining frequency classes A, B, C, D and E, frequency diagram for the area studied is prepared and compared to Raunkiaer's normal frequency diagram.

On the frequency figures, Raunkiaer propounded his law of frequency as follows :—

$$A > B > C \gtrless D < E$$

In addition to density and frequency classes, diameter or girth classes can be determined from the data obtained to give an accurate idea of space and cover of each species in a community. Thus space or cover classes can be formed.

The density, frequency, girth and space, or cover classes can be diagrammatically represented by phytographs. The phytographic representation of quantitative characters of plant communities provides a useful index for comparing two or more plant communities in a sere, or comprehending a change in the same plant community at some interval of time as shown in Fig. 13 (Rowe, 1949; Oosting, 1948).

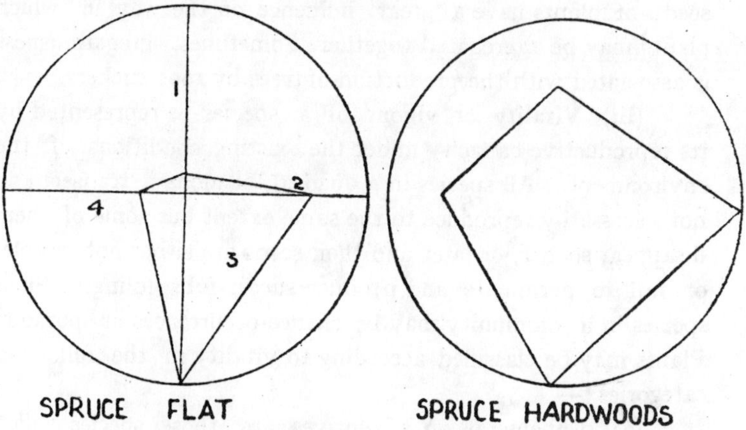

Fig. 13 :—Phytographs: radius 1, percentage of total dominant abundance ; radius 2, percentage frequency ; radius 3, percentage of total size classes represented; radius 4, percentage of total dominant basal area. (After Oosting, 1948).

QUALITATIVE CHARACTERS OF A PLANT COMMUNITY

While quantitative data of a plant community is being collected in quadrats, qualitative information on its important features at the same time can be recorded. This i nformation is collected under the following headings :—

(i) **Physiognomy** is a general appearance of the plant community and is described [l y stature of the plants, their spread and the characters of their life-forms. A grassland, scrubland, low forest, high forest, are some of the terms by which physiognomy over large areas is designated.

(ii) **Sociability, gregariousness or dispersion**—The way in which individuals of a species are grouped in a quadrat is called sociability, gregariousness or dispersion. Plants may occur, singly, in patches, colonies or in groups. Large colonies may form pure stands. Sociability is not a constant feature of a species but is generally determined by uniformity in the environmental factors and such community factors as competition, dependence, etc. Dispersal mechanism of seeds and fruits and the methods of reproduction—vegetative versus seed—of plants have a great influence on the way in which plants may be aggregated together. Sometimes, gregariousness is associated with the production of trees by root suckers.

(iii) Vitality or vigour of a species is represented by its reproductive capacity under the existing conditions of the environment. All species in a quadrat, or along a transect do not necessarily reproduce to the same extent but some of them disappear sooner, or later and their seeds are either not viable, or fail to germinate and produce successful seedlings. Such species in a community may be chance-occurrences or sporadic. Plants may be classified according to vitality in the following categories :—

(a) Ephemerals or adventives are those species which germinate occasionally and are not able to increase in number in a community. These are short-lived.

(b) In this category may be mentioned those plants which maintain themselves by vegetative means.

(c) Well developed plants, which regularly complete their life cycle and produce abundant viable seeds, belong to the third category.

Changes in vitality, or vigour of a plant species in a community give important indications to the rate and direction of succession. Species with high vitality become dominant or are potentially dominant and in the course of time determine the nature of the plant community.

(iv) Stratification—The way in which plants of different species are alligned in different layers or strata in a community is called stratification. Different life-forms may occur in different strata and indicate differences in physical and physiological requirements that plants in a community may have. Stratification in a community creates micro-environments and the growth and development of plants occurring in the lower strata are related to the physiological behaviour of species composing the upper strata. In ecological studies it is sometimes advantageous to represent changes in vegetation along a transect running through an area by stratification diagram or profile, one of which is indicated in Fig. 14.

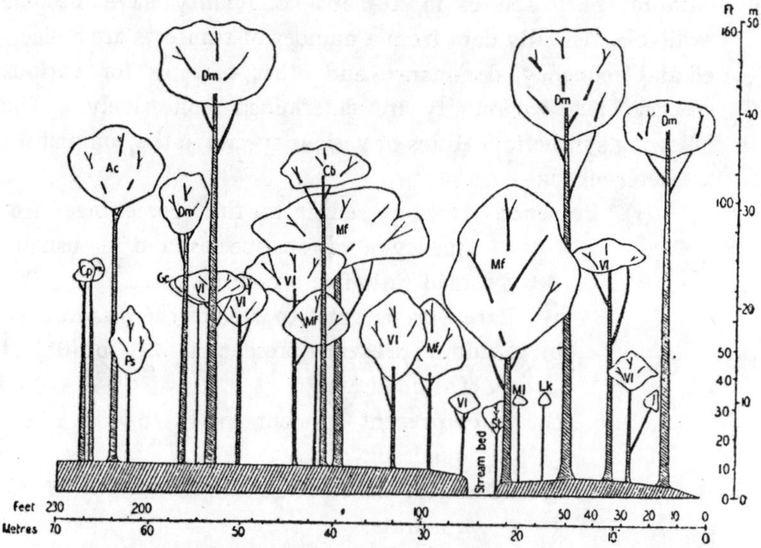

Fig. 14. Profile diagram of the *Dipterocarpus-Mesua-Michelia* community (By Srinivasan, from Rowntree, 1954)

Dm=*Dipterocarpus macrocarpus*
Cb=*Canarium bengalense* Co=*Chisocheton paniculaths*
Ac=*Artocarpus chaplasha* Ge=*Garcinia cowa*
Mf=*Mesua ferrea* Ps=*Polyalthia simiarum*
Vl=*Vatica lanceaefolia* Lk=*Litsaea khasiana*
Ml=*Myristica linifolia* St=*Saprosma ternatum*

(v) Periodicity. Seasonal or phenological changes in dominant species of a plant community are designated by the term periodicity. Seasonal aspects of a community may be vernal, aestival, autumnal or hibernal. Under monsoon climate in India rainy season and autumnal aspects are more pronounced. These aspects determine light intensity, moisture content and other features in the habitat and all micro-floral and micro-faunal activities on the forest floor are connected with seasonal aspect of the plant community.

SYNTHETIC CHARACTERS OF A PLANT COMMUNITY

These are detetmined after quantitative and qualitative data for each species in a plant community have become available. Usually data from a number of transects are collected and frequency, dominance and other features for various species in a community are determined collectively. The following synthetic features of various species in the community are determined :—

(i) Presence—is expressed from the percentage frequency of various species and is usually made into five presence classes.

(a) **Rare**—present in 1 to 20% of the quadrats.

(b) **Seldom present**– present in 21 to 40% of the quadrats.

(c) **Often present**—present in 41 to 60% of the quadrats.

(d) **Mostly present**—present in 61 to 80% of the quadrats.

(e) **Constantly present**—present in 81 to 100% of the quadrats.

The number of quadrats and transects necessary for determining the above feature is difficult to assess and may have to be worked out separately for each type of community. It may, however, be considered that in virgin or undisturbed forest a smaller number of quadrats may be sufficient, but in disturbed forest, or those subject to a scientific system of management the number of quadrats to be studied may have to be large. For comparative purposes it is, however, necessary that the number of quadrats in two communities to be compared must be the same.

(ii) **Constance**—If a species is present in 81 to 100% of the quadrats studied then it is said to be constantly present.

(iii) **Fidelity**—The degree with which a species is restricted in distribution to one kind of plant community is expressed by its faithfulness, or, fidelity. The folllowing fidelity classes are usually recognised :—

Fidelity 1—In this are included those plants which appear accidentally and are merely strangers in the community.

Fidelity 2—Plants, which are indifferent and occur in the community without any particular affinity with the community, have fidelity 2.

Fidelity 3—Preferents are those plants which occur in a number of plant communities but are predominant only in one.

Fidelity 4—Selectives—are a set of plants that are specially found in one community but may occasionally occur in other communities as well.

Fidelity 5—Exclusives plants are those which occur only in one particular plant community and do not occur in any other.

Plant species that have fidelity 3 to 5 are called characteristic species of the plant community.

Characterization of a Plant Community

With all the data having been collected on quantitative, qualitative and synthetic characters of the various species composing a plant community effort is made to select one or two species out of these, which would faithfully represent most of the characters of the community, and by which it would be possible to distinguish this community from all the others. Generally, a community is defined by one species, or sometimes by two; and in a few cases by three species. The species designating a plant community must be dominant, constant, with a high vitality, and fidelity. There is, however, a good deal of difference of opinion on the exact criteria on which species designating a community should be decided. In general practice, therefore, the characters given above are not strictly adhered to.

Correlation of Plant Communities with Environmental Features

After having characterised a plant community by its most dominant species and having determined its known features the next step is to attempt a correlation in space of the community with those features of the environment in which it lives. This is either done by relating a community with one or more characters of the climate, geology or soil, or by representing the environmental factor by some single value factor and relating it with the plant community. Actual correlation is shown by means of graphs, bar diagrams, profiles, or arranging the data with descending or ascending order in tabular forms.

Physiognomic Methods in the Study of Plant Communities:

Physiognomy is the study of the general appearance of plant communities. It depends upon the general stature, spread and life-forms of the species composing the community. There are a number of physiognomic methods which have been used in classifying plant communities; of these Raunkiaer's (1934) life-form method and Dansreau's (1951) methods are of some importance. The method of Schimper which was later on improved upon by Schimper and Von Faber is also of some importance as it laid down foundations of the physiognomic approach in the study of vegetation. Champion (1938) in India and Rosayro (1951) in Ceylon have classified vegetation following Schimper.

(1) Raunkiaer's life-form system. In the climax vegetation, or even in developmental communities dominant species are in equilibrium with their environment. The environment varies from place to place and the form or stature of plants in a community is usually governed by local variations. Such variations in plant stature are more prominently seen when the vegetation of large areas is considered.

Of all the factors of the environment oxygen, carbon dioxide etc., do not show much differences from place to place. Light intensity also does not vary greatly over large areas. Geology and soils, especially the physical and chemical nature of soil colloids differ very greatly, and, although these features are useful in determining small variations in plant life over small areas, it has not been found practicable to classify world vegetation into broad groups on these factors alone.

For classifying vegetation over large areas of the earth atmoshperic moisture and temperature were used by early workers. Raunkiaer considered that heat was probably of the greatest importance in determining plant distribution and he divided vegetation of the earth into the following four main classes on the heat requirement of species :—

(i) **Megatherms** require for their growth constant and high degree of heat throughout the year. They occur mainly in the tropics or in other areas where climate approaches tropical conditions.

(ii) **Mesotherms** can endure a considerably lower temperature during a longer or shorter period of the year, and are usually, but not always, found in tropical and subtropical habitats, and can stand competition with other plants in subtropical zones.

(iii) **Microtherms** are plants of temperate regions and demand much lower temperature for their growth and cannot withstand high summer temperatures. They may occur in tropical and substropical zones at high altitudes.

(iv) **Hekistotherms** are the plants of cold, alpine regions. These have the lowest heat demands and are short summer plants. They can endure long and very cold winters and are usually confined to arctic and alpine regions of the world.

Although, this classification on heat requirements of plants has been made, it is certainly not very satisfactory; for it has not been possible to name definitely, some structural features in a plant which may reflect its heat requirements.

It may be considered that heat determines the amount of moisture available for plant growth. Since water is of primary importance in the growth of plants it determines structural features by which it has been possible to say by looking at the plant to which moisture zone it may belong. The water economy of plants is dependent both on temperature and availability of moisture. Thus, drought may be caused by high heat and low moisture, or physiological drought may result even if high moisture is present, such as snow, when temperature is too low to make it available for plant growth.

In early geological periods when the earth was uniformly covered with thick vegetation from pole to pole heat and moisture conditions for the growth of vegetation over the entire world during the Carboniferous Periods were ideal. Later on, by the disruption of continents and by geological processes much of this vegetation was destroyed. The environment of plants changed and from the old vegetation there developed a number of new forms. These latter plant-forms differed from the older vegetation in one important aspect, that is in having semi-naked, or covered buds. The primitive vegetation which developed in uniformly warm and humid conditions had naked buds but species with covered buds and underground shoots originated later on, and these structures were developed in response to unfavourable conditions of the environment.

The present-day climate of most parts of the earth is characterised by seasonal differences occurring both in space and time. Thus excepting tropics where there is uniformly favourable season for plant growth, favourable and unfavourable seasons alternate in other regions. Plants outside tropics, therefore, adjust themselves during their growth to these two opposing sets of conditions. By the examination of temperature and precipitation curves for the whole year of a number of stations in different regions of the earth Raunkiaer found that unfavourable part of growing season, which is really the limiting factor, exerts a far greater influence on plant life than the favourable period. Therefore, adaptations in plant form, stature and function due to the limiting factor of heat and moisture are sound basis for classification.

On the location of buds in plants and the nature of the covering Raunkiaer classified plants of the earth into five broad categories. Life-forms is defined by Raunkiaer as "the sum of the adaptation of the plant to climate" and he considered that the plant climate can be expressed by the statistical distribution of life forms in the flora of the region.

1. **Phanerophytes** are those life forms in which the buds are naked or covered with scale, and are situated high up on the plant. These may be either evergreen or deciduous phanerophytes.

Depending upon the size of the plants, phanerophytes are divided into the following four main types :—

(a) **Mega-phanerophytes**—are trees which are over 30 meters (98.4 feet) high.

(b) **Meso-phanerophytes**—are trees which are 8-30 meters (26.2—98.4 feet) high.

(c) **Micro-phanerophytes**—are trees which are 2-8 meters (6.6 to 26.2 feet) high.

(d) **Nano-phanerophytes**—are shurbs which are under 2 meters (6.6 feet) high.

Phanerophytes are usually most abundant in tropical regions. Their number is greatly decreased progressively from sub-tropical to temperate climates. In temperate regions those phanerophytes are common in which the buds are protected and the plants are deciduous. In sub-tropical climates the majority of trees are of the evergreen phanerophytic type, with usually a bud—covering present.

The stature of the phanerophytes is best developed in tropical climates, but in sub-tropical and temperate regions they have low stature and show poor growth in girth also. In temperate regions micro-phanerophytes or nano-phanerophytes which are shrubby and low, occur. In snow covered arctic regions there may occur a certain proportion of nano-phanerophytes which are at the boundary between this group and the next life-form of plants, the chamaephytes. At very high altitudes and in desert climate only a very small number of phanerophytes exist.

(2) **Chamaephytes**–Chamaephytic climate occurs usually at high latitudes and high altitudes. In plants of this category surviving buds are situated close to the surface of the ground. In some types the aerial parts of the plants may usually die during the cold weather, leaving growing buds close to the surface. At the dawn of summer the plants again produce aerial shoots.

(3) **Hemicryptophytes**--Hemicryptophytic climates are found in the greater part of the cold temperate zone. In the plants of this zone surviving buds are actually hidden under the surface of the soil, usually protected by the soil itself, or by

dead remains of plants. The aerial parts of plants die back year after year. A large majority of the biennial and perennial herbs occurring in cold and temperate regions belong to this life form.

(4) **Cryptophytes or geophytes**—To this life form belong the plants in which buds are completely concealed in the ground. These are adapted to regions with a long dry period. Cold deserts and steppes have usually a large percentage of cryptophytes. Many herbs with horizontal root stocks and a majority of bulbs and tubers belong to this category of plants. During the favourable season when temperatures are high and moisture is available underground parts send up aerial shoots which quickly flower and complete their life cycle within a short period.

Hydrophytes are also cryptophytic plants, which have their buds below the surface of water.

(5) **Therophytes**—Therophytic climate is really a desert climate. Plants of this life form complete their life cycle in a single season and remain dormant throughout the unfavourable period in the form of seeds. They are abundant in hot and cold deserts.

BIOLOGICAL SPECTRUM

The percentage distribution of species among the life forms of a flora is called biological spectrum or phyto-climatic spectrum. By a study of all available phanerogamic flora of the world Raunkiaer made a normal spectrum, which he defined as "the percentage relations between the life forms of all the phanerogams of the world."

Biological spectrum is used to compare plant climates of the various parts of the world. This is done by comparing biological spectra of a region with the normal spectrum as prepared by Raunkiaer and classifying the climate on the basis of life forms. The life forms of a number of typical regions of the world are given in table 5.

The basic idea in Raunkiaer's life-form system is that the same life forms will occur in regions with similar evironmental conditions. To these regions he called equi-conditional

TABLE 5

Biological Spectra of different regions of the world.

Ph.=Phanerophytes; Ch.=Chamaephytes; H.=Hemicryptophytes; Cr. or G.=Cryptophytes or Geophytes; HH.=Hydrophytes; Th.=Therophytes)

Climate in terms of Raunkiaer's life-forms.	Region.	Schimper's and Von Faber's type.	Rubel's type.	Raunkiaer's life forms.					
				Ph.	Ch.	H.	Cr. or G.	HH.	Th.
	World Normal spectrum (1000 sp.) (Raunkiaer, 1934).	—	--	46	9	26	4	2	13
Ph.	Seychelles (Raunkiaer's, 1934).	1	1	**61**	6	12	3	2	16
	St. Thomas and S. Jan, Virgin Islands, (Raunkiaer, 1934).	1	1	**61**	12	9	3	1	14
	Sandy point, Danish W. Indies, (Raunkiaer, 1934).	1	1	**74**	16	4	1	0	5
Ch. & H.	Alps, Switzerland (Braun Blanquet, 1932).	14	9	0	**24.5**	**68**	4	--	3.5
	Spitzbergan (Braun Blanquet, 1932).	14	9	1	**22**	**60**	15	—	2
	Eastern Canadian Arctic (Desmaran, 1945).	14	9	0	**26.3**	**56.9**	11	13.7	— 2.0
H.	Central Switzerland (Braun Blanquet 1932).	5	7	10	5	**50**	15	—	20
	Paris area (Braun Blanquet, 1932).	5	7	8	6.5	**51.5**	25	—	9
	Connecticut (Ennis, 1928).	5	7	15	2	**49**	22	—	12
Th.	Death Valley (Braun Blanquet, 1932).	13	6	26	7	18	2	5	42
	Cyrenaica (Raunkiaer, 1934).	13	6	9	14	19	8	0	**50**
	Libyan Desert (Raunkiaer, 1934).	13	6	12	21	20	4	0	**42**

regions. It is very difficult to draw hard and fast boundaries between plant-climate regions on the basis of Raunkiaer's life forms, however, these provide a broad indication for the types of plants that are best suited for a particular region. Plant-climatic data could be of considerable application in selecting plants for afforestation and agricultural purposes and for comparing conditions in two or more plant communities. Two or more successional communities can also be compared and successional trends seen by life-form studies. It may be possible to apply this knowledge in silvicultural work for accelerating, or retarding the successional stream by planting or felling plants of a particular life form from a stand.

With a view to get additional data on climatic conditions of the region of which plant climate is determined by life-form enumeration Raunkiaer suggested the preparation of hydrotherm figures, with monthly temperature and rainfall curves. Hydrotherm figures for a number of places of which life forms are given in table 5 are shown in figure 15.

Biological spectra and hydrotherm figures have been given by Bharucha and others for a number of places in India and the plant climates for them have been determined. However, a word of caution is necessary in the use of such data. Since the figures relate to the flora and not to the bulk of the species growing in the region their social value is lost and the preponderance of one life form over the other has a very limited meaning. The biological spectra, moreover, generally indicate the most operative factor of the environment and not the climate. Thus, intense human exploitation may reduce a phanerophytic flora into a therophytic one and it is not safe to designate the plant climate in such a case as therophytic. Biological spectra may thus be a rough guide to the degree of biotic exploitation. Based on such assumptions, Pandeya (1954) in India, and a number of workers in America have estimated the extent of overgrazing in grasslands from a study of the biological spectra.

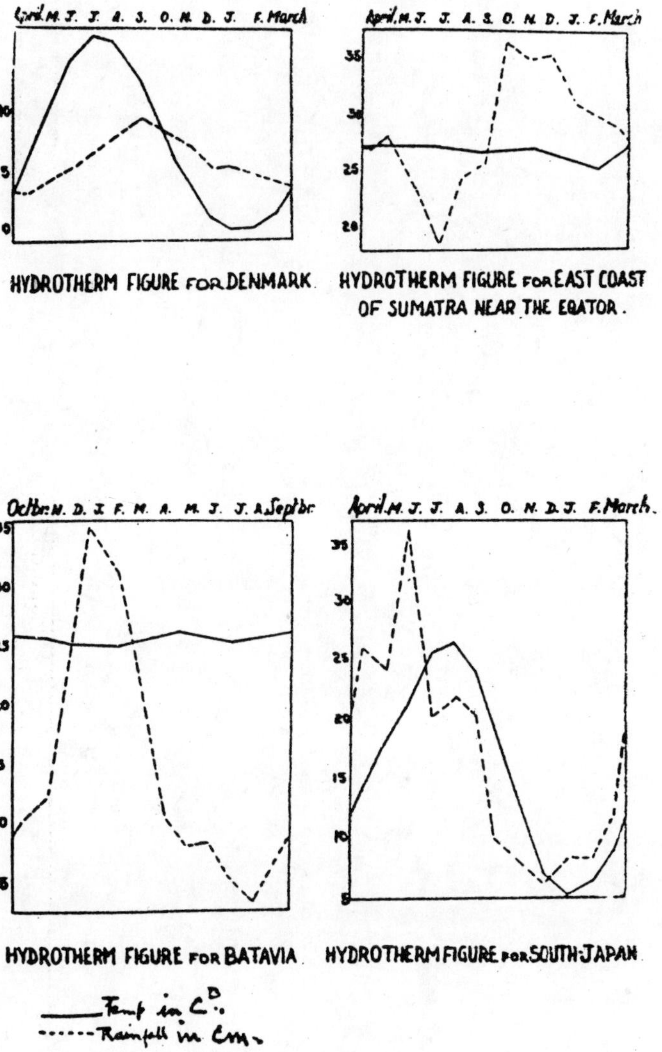

HYDROTHERM FIGURES. (After Raunkiaer.)

Fig. 15

Section B

Plant Environment

In this section environment of root (geology and soil) and shoot (climate) shall be described.

Plant Environment

The term environment etymologically means surroundings, and is analysed into a number of factors, such as moisture, temperature, soil, wind and gravity. For convenience of study environmental factors are grouped variously by different authors. Most of these have grouped them into three sets of factors, namely, climatic, edaphic and biotic. Daubenmire (1950) has, however, found it convenient to describe the environment in relation to vegetation under the following seven co-ordinate headings:—(1) Soil, (2) Water, (3) Temperature, (4) Light, (5) Atmosphere, (6) Fire and (7) the Biotic factor. Oosting (1948) envisages the environment as a complex of variable factors or causes, which includes:—

Substances as — Soil.
Water.
Conditions as — Temperature.
Light.
Forces as — Wind.
Gravity.
Organisms as — Plants.
Animals.
Man.
and Time.

There is complexity of interaction of all the factors in time and space which precludes sufficient justification in dealing them altogether separately. Perhaps a more cogent way of looking at them is to deal with the environments of the root and the shoot separately, since they are basically different. The various factors have, however, to be taken as discrete units strictly for the sake of convenience.

5

Environment of Roots

The environment of roots of terrestrial plants consists of the underground soil system which is a complex of interacting mineral particles, organic matter, aqueous solutions, air and micro-organisms. It is pertinent to understand the activities of the root system before proceeding to an analysis of its environment.

The Root System. Roots grow and ramify in between soil particles. During their growth they encounter mechanical resistance which can be overcome by pushing aside finer particles or by changing the route of growth. The growing tip has a protective root cap. Numerous branches arise and ramify and each one in its turn acquires a root cap, with the growing point behind from which a large number of the epidermal cells grow out into root hairs. The root hairs are in close contact with the finer particles of the soil. They are ephemeral and may not last longer than 48 hours, as in the meantime more root hairs arise from the advancing root tip. Thus, root hairs number into millions at any time on a root system and keep on engaging fresher particles as they are formed anew. The extent of contact of the root system with the soil is generally not fully realised. The surface area of the root system exposed to the soil system generally far exceeds the contact areas of the shoot and the atmosphere.

Busgen and Munch (1926) state "that the ability of pine to succeed where silver fir and spruce failed is the result of the pines possessing 24 times as many branches and root tips as the other species and 8 times the absorbing surface that they have."

Some of the records from Kramer of the length of roots and surface area they possess are as follows: A squash Vine sample from a green house had 15.9 miles of roots, and it was estimated that the root growth occurred at a rate of 1000 feet a day. A plant of wheat developed 44 miles of roots, a Rye 48 miles, and wild oats 54 miles. A crested wheat plant two year old was estimated to have 315 miles of roots and a winter Rye plant 4 months old possessed 387 miles of roots.

The root hairs absorb water and salts present in a film of solution around the particles and pass them on to the main root system for upward conduction. The whole soil system is concerned with the maintenance of the supply of these two substances and conditions for root growth.

It is necessary here to refer to the water and salt absorption by plants. Both the substances enter the root hair cell by the process of diffusion, according to the concentration gradient of each kind of molecule and ion. Water is the medium of salt movement but the two do not enter together in same concentration as they are in the soil solution. The diffusion pressure deficit and the selective action of the plasma membrane are responsible for uneven entry of various ions and molecules from a complex dynamic system of the soil solution. The metabolism of the plant is a specific character and each plant absorbs water and minerals according to its age, physiological status, species and their availability.

The soil and the vegetation are at any time in dynamic equilibrium so that the contact of the plant with the soil not only brings about a change in the plant alone but it also modifies the soil body. Both the soil and the plant are undergoing a continuous change which is dependent upon a number of common environmental factors. It is within human effort to change the nature of the soil, and influence plant growth, but it is comparatively much more difficult to change other environmental factors. Man was familiar with the relationship of plant and the soil since very early times.

Warming showed that special plants may occur on acid soils (oxylophytes), on saline soil (halophytes) on sand (psammophytes), on rocks (lithophytes), etc. The dominant role of soil in the growth of plants was recently shown more clearly by transplant experiments of the British Ecological Society and those of the Carnegie Institute. By growing the same set of plants, that grow in nature in different climatic conditions, on a number of soils transported at the same place it was shown that under the uniform conditions of climate soil exercises profound influence on the growth of different plant species. Plants in all stages of their growth, from germination to vegetative development, flowering and seeding are intimately governed by the edaphic factor.

Soil is the great source of moisture for the plants, and the role of soil in the growth of water plants which root in lake mud clearly demonstrates the superiority of the edaphic factor in plant environment. Plants have definite preferences for soils and are often used as soil indicators. There is a high correlation between vegetation types and soil types.

Stamp in Burma and Troup (1921) in India have shown that plants could be used for the classification of rock types in a general way. Many workers all over the world have similar experience in the study of forest vegetation. In regions of uniform climate soil is most often the controlling factor for plant growth and in regions with different climatic conditions the effect of the soil is no less important. For the ecological study of vegetation, therefore, soil is the most important single factor.

The Soil Complex. Ecologically, the soil is the medium for plant growth. It may be defined as the part of the crust of the earth in which roots of the plants are actually growing. In this way, bare crumbling rock, riverain or glacial deposit, bottom of a lake, solidified lava flow, etc., may all be considered as soils.

Biologically, soil may be considered as a weathered outer crust of the earth in which remains and products of decay of living organisms are finely mingled.

The pedological definition of the soil is given by Wadia (1945) as "the topmost layer of the earth's outer crust capping the rocks exposed at the surface. It is a natural body of variable thickness, composed of disintegrated rock material together with variable proportion of organic matter, mostly unconsolidated, generally differentiated into zones or layers, the lowest of which passes imperceptibly into the parent rocks below."

It is very often found convenient to visualise the soil complex as composed of the (i) mineral particle system (ii) the organic matter system, (iii) the biological system, (iv) the solution system and (v) the atmospheric system. All these systems are dynamic and involve actions and interactions among themselves and the factors of the environment of the shoot system leading to the development of the soil body into a definite pattern. These developmental patterns which are studied by means of soil profiles are distinctive for the general climate and vegetation of the regions. Variations may, however, arise by a multitude of factors of lesser importance.

Mineral Particle System

A cursory examination of any soil would show that it is an aggregation of particles of different sizes and shapes. The assortment of these particles differs from one soil type to another. Generally, three main groups of particle sizes are recognised (leaving aside gravel and grit which really speaking do not belong to the soil system). These are sand, silt and clay. They impart texture to the soil and according to preponderance of one kind of particle over others characterise it (Fig. 16). A classification of the particles according to the system of International Society of Soil Science is given in Table 6.

TABLE 6

Classification of soil particles according to system of International Society of Soil Science

(Partly after Lyon and Buckman)

Fraction	Diameter in millimeters	Sandy loam %	Loam %	Heavy clay %
Gravel or grit	10—1.0	...	...	...
Coarse sand	2.00—2.20	66.6	27.1	0.9
Fine sand	0 20—0.02	17.8	30.3	7.1
Silt	0.02—0.002	5.5	20.2	21.4
Clay	less than 0.002	8.5	19.3	65.8

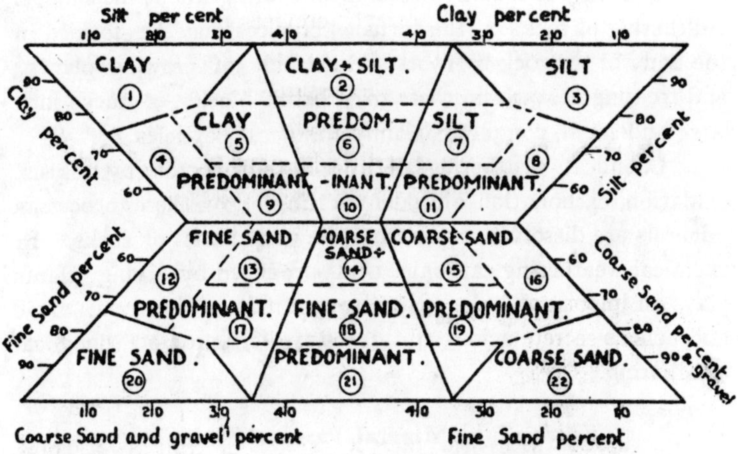

(1) Clay.
(2) Silty clay.
(2) Clayey silt.
(3) Silt.
(4) Coarse sandy clay.
(4) Sandy clay.
(4) Fine sandy clay.
(5) Loamy clay.
(6) Silty clay loam.
(6) Clayey silt loam.
(7) Loamy silt.
(8) Coarse loamy silt.
(8) Sandy silt.
(8) Fine sandy silt.
(9) Clayey loam.
(10) Loam.
(11) Silty loam.
(12) Clayey fine sand.
(12) Silty fine sand.
(13) Fine sandy loam.
(14) Sandy loam.
(14) Clayey sand.
(15) Coarse sandy loam.
(16) Clayey coarse sand.
(16) Silty coarse sand.
(17) Loamy fine sand.
(18) Loamy sand.
(18) Silty sand.
(19) Loamy coarse sand.
(20) Fine sand.
(21) Sand.
(22) Coarse sand.

Fig. 16.

Origin of the Mineral Particle System

The mineral particle system of soils takes birth by the weathering and disintegration of rocks by mechanical, chemical, or biological agents. Large diurnal, and/or seasonal variations in temperature and moisture conditions bring about mechanical weathering of rocks on the surface; crevices may be formed in the body of the rock by roots and rhizoids of growing plants; and freezing of water in these may break blocks of rocks into large and small, complex, or simple mineral particles.

Chemical weathering of rocks consists of hydrolysis, oxidation, carbonation and dehydration. By these processes minerals are dissolved away from the inert parts of rocks. In chemical weathering carbonic acid produced by living plants plays an important role. Numerous kinds of bacteria, algae and lichens secrete acidic substances, which promote chemical weathering.

Kinds of the Mineral Parental Material

The mineral particle complex may be residual or transported. The residual material of soils results from the disintegration of rocks *in situ*. Rocks usually weather on the surface, however, mass weathering of some types of rocks also takes place. The weathered layers usually contain most of the minerals which are present in the rocks. Beneath the weathered soil undecomposed rock is present.

(i) Residual Soils

Residual parent materials are related mainly to the geological nature of the rocks. The main types of rocks occurring in some important areas of India are given in Table 7.

The mineralogical and chemical composition of some of the important soil forming rocks of India are given in table 8.

The geological foundations of Indian soils are represented in the map (Fig. 17). Salient properties of the various zonal soil types were given by Wadia, Krishnan and Mukherjee (1945). The residual soil types are described below.

TABLE 7.

General Geological Succession in Different Parts of India

(Sy.=system; S.=series; B.=beds; Sst.=sandstone; Lst.=limestone),

Standard Scale	Northern Himalaya	Kashmir-Hazara	Simla-Garhwal	Assam	Peninsula	Coastal areas.
Recent	Alluvium	Alluvium & C.	Alluvium & C.	Alluvium, River gravels.	Alluvium Khadar	Alluvium sands
Pleistocene	Older alluvium, gravels, moraines,	Older alluvium, moraines, Karewa.	Older alluvium, moraines.	Older alluvium, gravels.	Older alluvium, Bhangar cave deposits, river terraces.	Raised beaches, Porbander stone etc.
Pliocene Upper Miocene	—	Siwalik system	Siwalik Sy.	Dihing S.	—	Siwalik Sy. Karikal B. Quilon B. Cuddalore Sst. etc.
Middle Miocene Lower Miocene	—	Muree S. Fatehjang zone.	Kasauli B. Dagshai B.	Tipam S. Surma S.	—	Gaj S.
Oligocene	—	—	—	Barail S. (in part)	—	Nari S. Dwarka B.
Eocene	Eocene	Chharat S. Ranikot S.	Subathu B.	Barail S. Jaintia S. Sylhet Lst. Disang S. Cherra Sst.		Kirthar S. Laki S. Ranikot S.
U. Cretaceous	Up. Cretaceous Chikkim S.	Up. Cretacecus	—	Disang S. (in part) Up. Cretaceous.	Deccan-Trap and Inter-trappeans Lametas,	Bagh B. Trichinopoly Cretaceous.
L. Cretaceous	Giumal Sst.	Giumal S. Orbitolina Lst.	—	—	Umia B.	Umia B.
Jurassic	Spiti Shales Kioto Lst.	Spiti shales jurassic	? Tal S.	—	Jubblepore Kota Rajmahal	Katrol Chari Patcham
Triassic.	Triassic	Triassic	? Shali Lst. ? Krol S.	—	Maleri Pachmarhi Panchet.	—
Permian.	Productus B. Syringo-thyris Lst.	Agglomeratic Slates Zewan B.	? Infra-krol.	—	Raniganj S. Barren Measures, Barakar S.	—
U. Carboniferous	—	Agglomeratic sh. Gangamopteris B. Tanakki Boulder B	? Blaini boulder Bed.	Subansiri B.	Umaria B. Talchir boulder B.	—
M. & L. Carboniferious	Po seried Fenestella sh.	Fenestella sh.	—	—	—	—
Devonian	Muth Qtzt.	Muth qtzt. Devonian (Chitral).	? Jaunsar S.	—	—	—
Silurian	Silurian	Silurian.	—	—	—	—
Ordovician	Ordovician	Ordovician.	—	—	—	—
Cambrian	Cambrian	Cambrian	? Deoban Lst.	—	Vindhyan Sy.	—
Algonkian	Haimanta Sy.	Dogra slates	Simla slates.	Buxa S. Daling S. Darjeeling S.	Cuddapah Sy. Delhi Sy.	—
Archaean.	Vaikrita Sy.	Salkhala S. Gneisses.	Chail S. Jutogh S.	Shillong S., Gneisses.	Dharwar Sy. Aravalli Sy. etc. Gneisses.	—

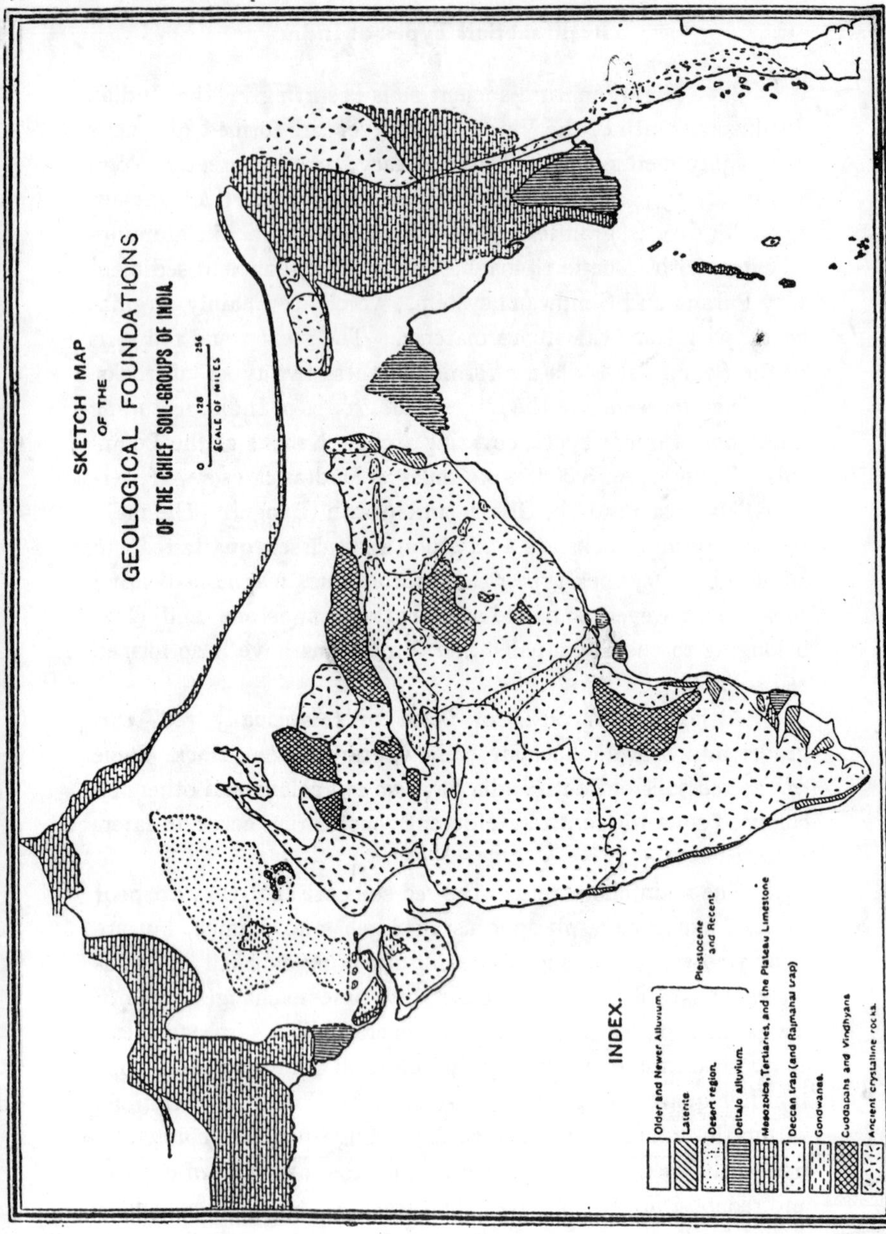
SKETCH MAP
OF THE
GEOLOGICAL FOUNDATIONS
OF THE CHIEF SOIL-GROUPS OF INDIA.
0 128 256
SCALE OF MILES
INDEX.
Older and Newer Alluvium.
Laterite.
Desert region.
Deltaic alluvium.
Pleistocene and Recent
Mesozoics, Tertiaries, and the Plateau Limestone
Deccan trap (and Rajmahal trap)
Gondwanas.
Cuddapahs and Vindhyans
Ancient crystalline rocks.

Fig. 17.

Residual Soil Types of India

(a) Highly mature ancient soils occurring in the Indian Peninsula, south of the Vindhyan Range, are formed of gneiss and highly metamorphosed sedimentary rocks of great geological antiquity. The materials of these soils consist of ancient crystalline rocks, granites and gneisses which are acidic alumino-silicates, with some ferro-magnesium salts, and ancient sedimentary Purana and Gondwana systems, which are mainly argillaceous, with some calcareous material. The great zonal red soils of the Peninsular India are formed of these two types of rocks.

The red soils are the largest soil groups of India, including a number of minor types, covering archaean rocks of the Peninsula. These cover 8,00,000 sq. miles in Madras, Mysore, Hyderabad, Madhya Pradesh, Orissa and South Bengal. The main red soil-forming rocks are acid granites, gneisses, quartzites and some subsidiary rock types, rich in iron and magnesia-bearing minerals, however, some old sedimentary sandstone and clays belonging to the Cuddapahs and Vindhyans have also formed red soils.

Although, the colour of these soils is usually red, this grades into brown, chocolate, yellow, gray or even black. Some of the red soils of South India have no red colour and other red coloured soils with entirely different constitution occur in lateritic areas.

The main characters of the red soils are that they are poor in lime, magnesium, phosphorus, nitrogen and humus, but are rich in potash. These are generally speaking deficient in soluble exchangeable bases, and their total base exchange capacity for potassium, calcium, etc. is generally low. In chemical composition these are mainly siliceous and aluminous and contain free quartz and sand. There are a good deal of variations within this zonal group of soils, depending upon the configuration of the area, drainage pattern and the amount of erosion and redeposition which have taken place. The drainage conditions have determined the intensity of leaching and the accumulation of salts at lower depths in the soil profile. Sub-types

of this zonal group have been made according to the presence or absence of calcareous *kankar*, or *ferruginous iron pan*.

A good deal of work on the chemical composition and agricultural properties of the red soils has been done by Ray-Chaudhuri and others (1937, 1941, 1943), but very little is known of the forest areas. These bear evergreen forests of the coastal areas and in the western ghats there is a scrub vegetation, which has been exploited badly. Some of these areas have been clear-felled to raise plantations of teak, which have not been much success.

(b) The Deccan plateau basalts are composed of basic ferro-magnesian lavas and ash-beds. They form *black soils*, or *regur* of the Peninsular India. The black soils again include a number of distinct varieties, and types, but the term black soils is applied to groups of soils, occurring in the Deccan, Bombay, Malwa, Hyderabad, Gujerat, Kathiawar, Madhya Pradesh and Berar. The most characteristic black soils cap the basaltic plateau of the Deccan traps and are argillaceous, fine-grained, dark coloured, clayey and contain high contents of iron, calcium, magnesium and aluminium. These are poor in phosphorous and nitrogen. Potassium may be present in variable amounts and calcium carbonate and magnesium carbonate are abundant. These soils in the valleys may be of considerable depth and are differentiated into various horizons.

Although, majority of the black soils occur on the Deccan trap, they may also be found on the adjoining granite, basic gneiss, sandstones, schists of the Vindhyas and Cuddapahs. The typical black soil popularly called *black cotton soil* has a high clay fraction (40%—60%), without gravel or coarse sand. It retains good amounts of moisture and becomes very sticky when wet; on drying, large and deep cracks are formed. The black cotton soils of Gujerat and Berar have high amounts of calcium carbonates (6% to 8%), iron oxide (9% to 10%) and a fairly constant amount of aluminium (10%). Potash is usually less than 5% and phosphates, nitrogen and humus are, as a rule, low. The *regurs* are generally highly productive, but on uplands they are not so. The base exchange capacity of the black cotton soil is generally high throughout the profile.

The cause of the black colour of the soil has been a subject of much controversy. Sigmond (1938) believed it to be due to humus; Annet (1910) thought it to be due to finely divided titaniferous magnetite. A number of other pedologists think that the black colour arose due to colloidal iron and aluminium silicates.

The black cotton soils bear a mixed deciduous forest, in which teak predominates. The common associates of teak are *Terminalias, Anogeissus, Diospyros,* etc. On somewhat water-logged areas *Butea monosperma* is found. On riverain black dposits *Acacia arabica* and *A. leucophloea* are common species.

In dry patches of black alluvia grasses and bamboos are most common. In the Nilgiris, black cotton soils occur in narrow valleys in which due to water-logging and temperate conditions at high altitude peaty conditions develop. Some of those areas are being drained and planted with willows.

(c) Laterites are of high and low level types. These are surface deposits and are formed by the sub-aerial alteration of rocks. They are peculiar to India and some other tropical countries. In India, these occur in the Peninsula and some parts of Assam and also in Burma.

Although opinions differ, it is held that laterites have been formed during the Pleistocene and Pliocene periods under humid tropical conditions. Humic acid is rapidly oxidised on formation, as a result of which silica is leached out in preference to sesquioxides. In this way, soils become rich in hydrated oxides of aluminium or iron and due to the presence of red coloured iron oxide the soils acquire red colour.

In some cases, aluminium oxide is more than 50% in the rock and this rock is called aluminous laterite or bauxite. Some studies in laterite and lateritic soils of India have been done by Fox (1936) and Ray Chaudhuri and his co-workers (1937, 1941).

On account of their colour, soils derived from laterite are often classed with red soils. But true laterite soils differ from the red soils in being largely composed of hydrated sesquioxides of alumina and iron with free silica, in the form of quartz grains. Phosphoric acid and titanium are present, but alkalies and alkaline earths are absent, especially calcium and magnesium salts. These are acidic and contain little, or no humus, and SiO_2/Al_2O_3 ratio is low in a mature sample. On account of their having been intensely leached during their formation, laterite soils are generally very infertile. At lower levels due to the admixture of other transported material they are used for agriculture, but at high levels these soils are thin, and gravelly and generally do not retain much moisture.

In some parts of the country, for example, in the Nilgiris and Madras, a number of red earths and clays have been wrongly grouped under laterite soils. Unlike true laterites these red soils are generally rich in silica and lime and have a high ratio of SiO_2/Al_2O_3, and they do not contain colloids, which are essentially present in laterites. These should best be called lateritic soils.

Laterite and lateritic soils occur in Australia, some parts of Europe and in North America. Stephens (1949) has done some work on comparative morphology and genetic relationship of these soils. He believes with Prescott that laterite in Australia is a fossil soil. In that country, they are truncated in varying degrees, and it is considered that they are associated either with fluctuating water table, or they are of a relict, or fossil character.

The laterites in south India usually bear an evergreen forest, in which chief tree is *Xylia xylocarpa.*

In the Peninsular India some soils are associated with sandstones, slates and limestones of the Cuddapah, Kurnool and Vindhyan systems. These rocks have only a thin residual cap, which is mainly composed of insoluble ingredients. Being ancient geologically, these soils are highly mature and seem to have attained equilibrium with the factors of the locality. The soils on the limestone are thin, and depending upon the impurities which were present in the rock originally, these are either argillaceous, or arenaceous. These usually contain fair amounts of exchangeable bases and are sometimes alkaline in reaction, but they are, curiously, often deficient in lime.

A type of thin and immature soil of low fertility occurs on the Gondwana rocks. These soils are low in exchangeable bases and being sandy in composition are either barren, or support only a dry type of deciduous forest.

(ii) Transported Mineral Particles.

Transported soils are classified into various types depending upon their mode of transportation. Those transported by gravity are called *colluvial.* Such soils from *talus, scree-cones, landslides, mud-flows,* etc., in mountainous regions. Colluvial deposits are characterised by having no stratification, and in these the particles of the soil are of many sizes and shapes, which are not sorted out. Of the various types of colluvial deposits landsildes and mud-flows are formed suddenly. Most colluvial deposits on the Western Himalayas on stabilisation bear *chir* pine community with numerous other plants. Blue pine may also be found on colluvial deposits of schists, or phyllites.

The soils transported by running water are called *alluvial.* Alluvial deposits occur in the from of *out-wash-plains, flood-plains, terraces* and *fans*. These deposits are characterised by having rounded and smoothed particles of soil, which are generally well stratified. Depending upon the movement of water alluvial deposits show strata of finely-grounded silt, clay or sand. They exhibit boulders or large blocks of rocks in their body. The flood plain deposit in the Himalayas bear different types of forest communities. In the Kulu valley they show the following succession.—*Chir*-pine→Blue pine→ Deodar. If there is no biotic interference the *Quercus incana* climax is reached. In the plains, the succession is from *Acacia catechu*→*Dalbergia sissoo*→*Holoptelea-Trewia*→a mixed deciduous forest.

Wind blown deposits :—Among the pioneers which bind the sand on coastal areas are species of *Spinifex, Ipomea,* etc. On stabilisation *Memecylon edule, Calotropis gigantea, Eugenia jambolana* and palms form various types of plant communities.

On inland sand dunes as in Rajputana the details of succession are different. Large scale wind transportation of sands from the Rajputana desert towards the east is extending its area gradually in an arc. In order to arrest the march of the desert towards the Gangetic plain a four mile deep belt of vegetation, nearly 200 miles long, is being planted with *Prosopis* and numerous other species. The work is in an experimental stage.

Another kind of wind blown deposit, which occurs in some parts of northern India, is *loess*. This material has been deposited in the post-glacial period, and forms thick deposits in the Potwar plateau in Pakistan. The material of *loess* is very fine and contains large amounts of organic matter. This is covered by various species of *Acacia*.

Glacial Deposits

Soils transported by glaciers are called glacial deposits. Glacial deposits which may also be called *drift* or *till* are recognised by their having angular boulders and particles of different sizes, which do not exhibit any stratification, but lie in a pell mell condition. *Glacial mor aines* and *out-wash* deposits of the Pleistocene period occur extensively in the Himalayas. Moraines of three main glacial advances have been recognised in the Kashmir Himalayas. Morainic deposits have different composition, depending upon the type of rocks from which the boulders and matrix are derived. These are usually highly fertile and are covered by extensive forests of conifer species (usually *Abies webbiana* also *Pinus excelsa*) and broad-leaved temperate plants, like birch, *Prunus*, poplar, etc. On scree deposits at high altitudes *Rhododendrons*, and many cushioned plants, dwarf willows, junipers, etc. are present.

Indo-Gangetic Alluvium:—These soils are old transported silts, clays and sands with concretions of calcareous matter at various depths. The Indo-Gangetic belt of alluvial soils extends from Sind, Rajputana, greater part of the Punjab,

U.P., Bengal, Bihar and Assam. The parent material of these soils is entirely of transported origin and these are either azonal, or have developed very weak soil profiles. These profiles should be considered as distinct from the usual profiles in forests of temperate countries.

Depending upon the climate, rainfall, humidity, aridity, wind action, etc., these soils have developed into characteristic sub-types. On the whole, they contain low amount of nitrogen; but alkalies, potassium and phosphoric acid are usually adequate. Lime and iron oxide greatly vary in these soils. The exchangeable bases are sufficient, and mineral plant foods are fairly large, and on account of their being easily replenished these soils have formed good agricultural areas. They are divisible into two natural groups. The younger *khadar* soils of the North-west India are drier, pale-brown, sandy clays and loams, with small amounts of calcareous matter. The most common plants on them are *Dalbergia sissoo, Acacia arabica, Morus alba, Salvadora oleoides, Capparis aphylla,* etc. The common grasses are *Saccharum munja* and *S. spontaneum.*

The other, a little older soils are *Bhangar,* which occur in the Gangetic Basin of Oudh, Bengal and Bihar. These are darker-brown, argillaceous loams with a larger clay content and less sand. These are more moist and have high content of calcium with nodular *kankar.* In these soils water table is relatively higher than in the *khadar* areas. These bear a good type of *Shorea robusta* forest.

At some places due to the presence of impervious layers of clay and hard pans drainage becomes impeded at various depths. The defective drainage promotes the accumulation on the surface of the soil profile of sodium and magnesium salts, which are injurious to many plants.

Reh, usar and *kallar* soils which are impregnated with salts and are highly alkaline are present in drier parts of Rajputana, Punjab, U. P. and Bihar. These soils are notably impervious. liberate larger amounts of sodium, magnesium and calcium salts and sulphurous acid on weathering. These are largely infertile areas bearing extensive growth of grasses, with little or no tree growth. *Butea monosperma* is the only conspicuous plant in the area. The afforestation of these areas is engaging the attention of the Central and State Forest Departments.

The introduction of canal irrigation in the Punjab and West U. P. had resulted in the production of *usar* and *kallar* soils. These soils are usually low in nitrogen and organic plant food material. In *reh* soils, sodium carbonate and sulphate and calcium and magnesium compounds are most commonly present. In some *usar* and *kallar* soils in the Punjab *Sueda fruticosa* and *Kochia indica* are common plants.

Considerable amount of work on the soils of the Gangetic basin is being done by the new regional soil laboratories set up by the Agriculture department of U. P., and the Central Soil Survey department.

Delta Soils. Soils occuring in the deltas of the Indus and the Ganges are loamy and rich in organic matter. They contain high amounts of moisture due to defective drainage, and being periodically flooded, are generally very fertile. At some places, marshes, swamps and jheels are formed and due to alternate flooding and drying a glei horizon develops in such places. The Sunderbans in Bengal constitute one of the most important examples of the delta soils of India. On fresh water delta soils the predominant tree is *Heritiera minor*. On river banks and alluvia *Sonneratia apetala* and *Nipa fruiticans* are found. On salt water forests *Bruguiera gymnorrhiza* is occasionally found. On river banks in the area *Rhizophora conjugata, Kandelia, Avicennia,* etc. are found.

A large tract of dry, sandy, barren waste derived largely from the mechanical disintegration of ground rocks, and sand blown from sea coast is present in regions of very low rainfall in western parts of India. At some places, these soils at the surface are heavily impregnated with sodium chloride and alkaline salt. They are deficient in clay factor and form barren sandy soils. Examples of these may be Thar and Rajputana desert. *Sesbania* spp., *Prosopis* spp., *Crotalaria, Tephrosia, Leptadenia, Calotropis,* etc. occur commonly in the area. The afforestation of these deserts is actively engaging the attention of the Union Government.

Two other minor alluvial soil types, namely, *Bhabar* and *Terai,* which form transition into the Indo-Gangetic alluvial soil, occur in the northern parts of U. P. The *Bhabar* are chiefly composed of gravel and shingle beds, supporting a thick forest of *Shorea robusta, Anogeissus, Bauhinia, Terminalias,* etc. The *Terai* areas form bog and marsh soils that have usually a high grassy growth. Some of the *Terai* areas have recently been reclaimed and put under agriculture.

(iii) Mountain Soils.

Pre-Tertiary and Tertiary sedimentary systems of extra Peninsular mountains consisting of sandstones, clays, shales and limestones are present on the Himalayas. The lower part of these rocks is more calcareous, while the upper is more sandy and clayey. Large masses of intrusive granites of ancient age occur in inner dry Himalayas. Mountain and forest soils of the Himalayan region are usually found in depressions and valley basins, or on gently inclined slopes. The southern faces of these mountains in most cases being scarp slopes have a very thin layer of the soil, while deep profiles develop on the northern, dip slopes, which are moist and cool. The main forest types in the North-West Himalayas on the dip slope are mixed conifers; however, oaks and other deciduous broad leaved species form pure or mixed belts in the conifer forests, predominantly on scarp slopes. Some amount of work on forest soils of the Kashmir and Kulu Himalayas was done by the Irrigation Department of the Punjab and recently these studies were extended to Bashahr Himalayas. There are no oaks in the Kashmir valley due to glaciation in the Pleistocene period. In un-glacial areas, oaks predominate, however. Two main soil types resembling somewhat north European podsols and brown earths have been described. Some studies in the oak forests of the Eastern Himalayas were also recently done. These soils are far more acidic than the West Himalayan soils and in this respect resemble the north European podsolic soils (Puri, 1954). These bear mixed oak—laurel forests and recently large areas have been planted with *Cryptomeria japonica.*

In the Siwalik foothills region that extends from Afghanistan to Assam, Tertiary sandstone and clay stone form primary soils. These are shallow and azonal, and contain a large proportion of undecomposed mineral grains. They are sandy, or gravelly, are usually porous and impregnated with lime and sodium salts. These do not contain much humus. In their northern parts some Siwalik soils resemble mountain forest soils in showing podsolic and brown earth characteristics in their profiles. The characteristic vegetation in the Siwaliks consists of *chir* pine, *sal* and other species that are found in the *Bhabar* areas.

Even at very high altitudes in the Himalayas true podsols have not so far been found (Taylor *et. al.*, 1939). Wadia (1945) records mountain meadow and highland steppe soils from regions above snow line. Above the altitude of about 14,000 ft. the soils are generally frozen for the greater part of the year and on account of the intense leaching by ground water they are thin and have podsolised profiles.

The last Ice age in the Himalayas during the Pleistocene has left large glacial deposits of moraines, glacial outwash; and at some places varved clay is also present (Puri, 1948). These are covered with thick forests and on account of their unstable conditions large landslides, slips and creeps occur in these. These are, however, very fertile.

Some of the deep, clayey mountain soils with high amount of humus are maintained in central and eastern Himalayas as tea gardens.

In the Nilgiris, shola temperate forests and grasslands are found. In hills of Central India forests of teak, *Shorea robusta* or of mixed deciduous types are found, depending upon the nature and character of the mineral particle system and others factors of the environment.

Soil Map Of India

Some work has been done on a classification of chief soil types of India and a preliminary soil map (Fig. 18) was prepared by the Indian Council of Agricultural Research in 1943. In this work only visual characters of soils have been considered as criteria of classification.

Recently, this map has been revised and a better soil map of India (Fig. 19) has been presented. In this map, although the basis of classification is not much different from the first soil map, additional data to correlate soil types with major climatic zones have been given. This is definitely an improvement on the older map.

Raychaudhuri has recently drawn up a basis on which further work on the preparation of a soil map of India should be done. He considers that data on soil profiles, morphological and chemical characteristics of soil, together with vegetation ought to be collected.

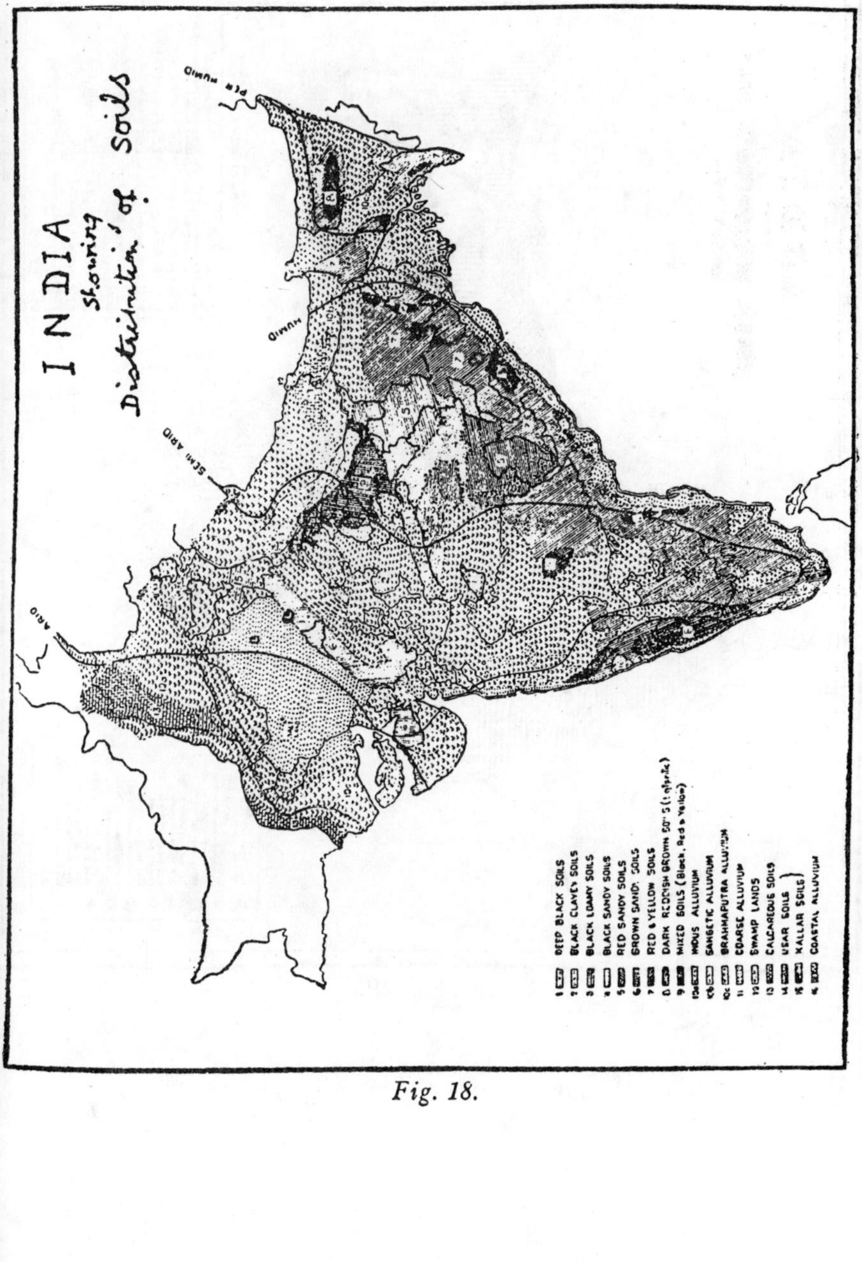

Fig —Showing the distribution of the various soils of India (Reduced from the Soil Map of the Imperial Agricultural Institute, New Delhi)

Fig. 18.

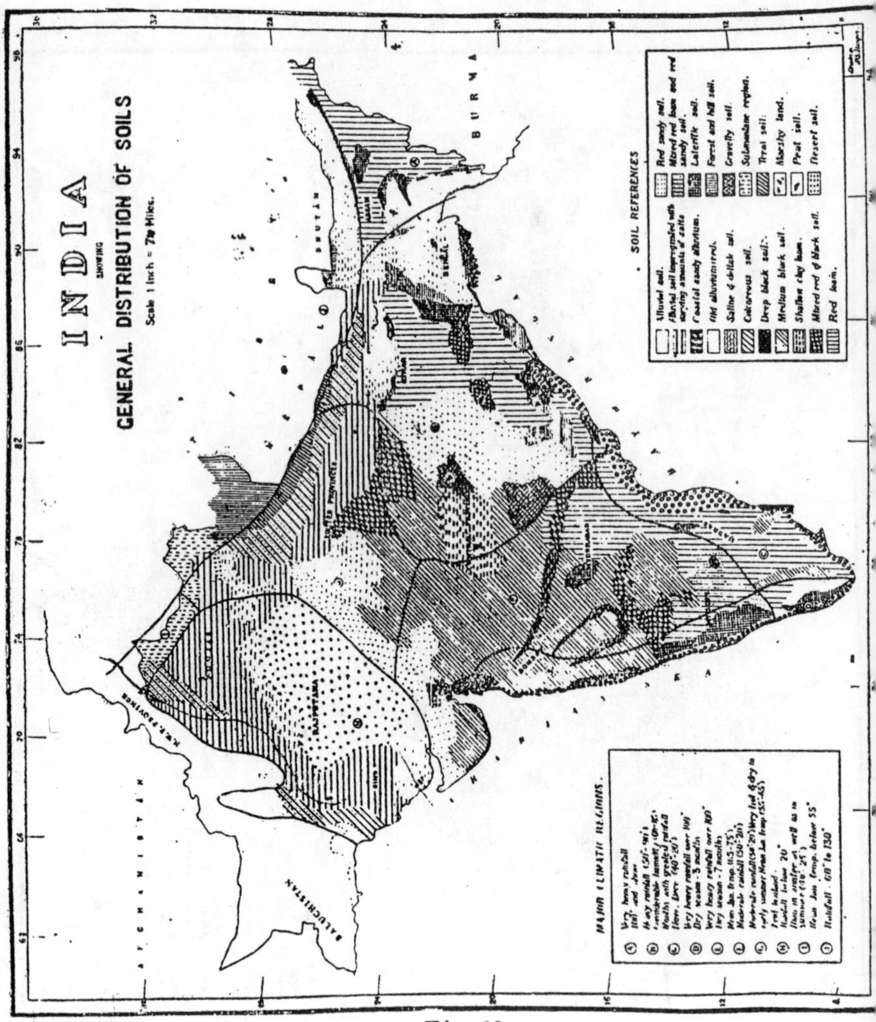
INDIA
SHOWING
GENERAL DISTRIBUTION OF SOILS
SOIL REFERENCES
Alluvial soil.
Alluvial soil impregnated with varying amounts of salts
Coastal sandy alluvium.
Old alluvium.
Saline & deltaic soil.
Calcareous soil.
Deep black soil.
Medium black soil.
Shallow clay loam.
Mixed red & black soil.
Red loam.
Red sandy soil.
Mixed red loam and red sandy soil.
Laterite soil.
Forest and hill soil.
Gravelly soil.
Submontane region.
Terai soil.
Marshy land.
Peat soil.
Desert soil.
MAJOR CLIMATIC REGIONS
AFGHANISTAN
BALUCHISTAN
RAJPUTANA
BURMA
BHUTAN

Fig. 19.

6

Biological system of the soil

The soil is not a lifeless body. A very large variety of organisms live in it, and form a vital link in its development and supply of nutrients to the root. Bacteria and protozoa in millions and fungal hyphae in great abundance are busy decomposing organic matter. Earthworms, rodents, insects and other larger burrowing animals aerate the soil. The roots themselves are highly reactive on their environment—removing salts and water and yielding chemicals of complex nature and organic matter on their death. The respiration of these organisms is responsible for depletion of oxygen and addition of carbon dioxide to the soil atmosphere.

The role of organisms in forest floor.

Soon after its reaching the forest floor tree litter is acted upon by atmospheric oxygen and rain water. Water soluble substances begin to be washed out and the organic material is attacked by a large variety of soil population, among which bacteria, fungi, worms, insects (especially in temperate region), termites and ants (in tropical countries), predominate. These organisms feed upon this dead material and as shown by Charles Darwin and later confirmed by numerous workers, litter rich in calcium and nitrogen is preferred by worms. Worms are usually more obvious in the surface layers of broad leaved deciduous forests of temperate regions and as Bournebusch (1930) had shown in his study of Danish forest soils their numbers in conifer forests is very small. Puri's (1950) studies in English woodlands have shown that worms are more obvious in mull soils or mull-like-mor soils, which are less acidic, have a normal relative humidity, and organic matter between 7% to 8%. In Indian forest soils worms become evident only during the rainy season. During the dry season they perhaps burrow deep into the soil. The Indian soils although analogous to mull type differ widely from it in the activities of worms.

Worms not only devour a large amount of forest litter but as Salisbury (1924), Wherry (1924), Puri (1950) and others have shown the worm casts thrown on surface of the soil are rich in organic matter, have a higher amount of water soluble bases and nitrogen, and lower acidity than the neighbouring soils from which they are derived.

The favourite food of worms in temperate forests consists of leaves of ash, poplar, hazel, lime, etc. which are generally rich in calcium and nitrogen.

It is estimated that nearly 6 to 25 per cent of the total litter production in a beech wood is consumed by worms. The population of worms in a normal clayey loam with a mull type of humus layer is usually very large and is estimated to be of the order of 5,22,000 to 10,10,000 per acre in some coarse mull soils of New York. The number of worms in per sq. meter area in mixed hardwoods of Sweden is estimated to be 150.

In tropical countries termites and ants have a very large population and their effect on the forest litter is not adequately known, although, it is recorded that some soils of termite mounds in some Indian forests contain a higher amount of bases and nitrogen than the neighbouring soil. Besides these there is a great population of micro-fauna and micro-flora the estimates of which in some pine, fir, redwood, juniper and spruce forest soils in the United States are given in Table 9 Jensen (1931) has given some interesting estimations of fungi and bacteria in different types of soils. In Table 10 it will be seen that the number of those micro-organisms varies a good deal with the type of soil. Similar estimates for Indian

TABLE 9

Micro-organisms in forest soil profiles from some pine, fir, redwood, juniper and spruce forests of the United States.
(Modified from Powers and Bollen, 1935)

Sample	H_2O free basis.	Numbers per gram, water free basis Population in millions of Bacteria	Actino-mycetes	Molds.
1. De Kalb silt loam air dry samples of 4/23/32 incubated at 28° and approximately 1/2 saturation for two weeks 6/14/33.				
Litter	566	4.32	2.52	2.39
F-layer	344	4.88	none	7.39
H-layer	212	9.57	none	13.20
*A	23.5	1.41	none	1.00
2. Olympic silt loam-field condition, not incubated.				
Litter	25. 0	66.78	0.88	0.10
F	22. 3	47.53	11.62	0.08
H	223	34.54	9.23	0.17
H, transition ...	147	7.62	4.73	0.06
A_1	6. 9	1.98	0.64	0.03
A_2	31. 9	4.32	0.73	0.01
B	38. 9	0.59	0.07	very few
C	63. 4	0.17	0.08	very few

* No Azotobacter found in A.

TABLE 10

Numbers of Fungi, Bacteria and Actinomycetes in soils of different characters.

(Modified from Jensen, 1931).

No.	Character of soil	pH	Fungi per gram thousands	Bacteria+ Actinomycetes per gram (millions)
	Field soils			
5	Sand	6.59	218.0	12.0
16	Loam	8.86	80.0	16.7
22	Heavy loam	7.52	78.0	16.2
38	Clay	4.96	98.0	1.75
41	Humus	7.08	174.0	25.9
	Garden soils			
52	Light loam	7.18	154.0	—
58	Light loam	7.78	173.2	25.7
	Forest soils			
59	Sand	4.97	229.2	4.55
60	Sand	5.12	78.0	0.83
62	Humus ('Mould')	5.94	45.6	—
64	Humus ('Mould')	4.12	165.0	0.67
66	Humus (Peat)	4.32	82.0	0.61
69	Humus (Peat)	4.53	56.0	—
	Moor and meadow soils			
77	High moor	4.58	220.6	1.06
80	Low moor	5.20	108.0	3.55
85	Mud	6.90	68.6	23.3
	Heath soils			
86	Dry peat	4.40	40.4	—
88	Dry peat	4.10	185.4	0.40
90	Dry peat	4.39	127.4	1.23
	Salt marsh soils			
91	Very heavy clay	6.82	29.7	8.0
94	Heavy clay	6.29	31.0	—

forest and grassland soils are given in Table 11 after Saksena (1955).

TABLE 11

Fungal population in the soils of Sagar, M. P.
(After Saksena, 1955).

Situation	Vegetation (3 most frequent spp.).	Soil depth	No.* of fungi per gm. of soil.
Plateau top	*Anogeissus latifolia*	6″	46,890
	Nyctanthes arbor-tristis	12″	7,490
	Tectona grandis	18″	5,450
Mid slope	*Butea monosperma*	6″	56,280
	Diospyros melanoxylon	12″	16,600
	Anona squamosa	18″	2,290
Base of hill	*Tectona grandis*	6″	39,990
	Diospyros melanoxylon	12″	15,540
	Butea monosperma	18″	3,790
Ghatera forest	*Tectona grandis*	6″	54,920
	Butea monosperma	12″	9,530
	Acacia leucophloea	18″	5,810
Grassland	*Dichanthium annulatum*	6″	70,520
	Bothriochloa pertusa	12″	26,160
	Iseilema anthephoroides	18″	9,760

* No. estimated by plate count method.

Animals play an important role in the decomposition of litter and formation of humus. A comparison in the animal population of Douglas fir-hemlock community and oak community in some American forests is given in Tables 12 and 13.

TABLE 12

Abundance of soil forming fauna of the given forest communities in the U. S. A.

(After Dirks Edmunds)

	Douglas fir-hemlock community.	Oak community.
Invertebrates. (Based on M² collections).		
Shrub layer, average number of individuals ...	134	123
Herb layer, average number of individuals ...	270	199
Ground layer, average number of individuals ...	559	1251
Total ...	963	1573
Insects:- other than *Apterygota*: (Based on M² collections).		
Shrub layer, average number of individuals ...	74	117
Herb layer, average number of individuals ...	257	187
Ground layer, average number of individuals ...	134	519
Total ...	465	822

TABLE 12 (*Continned*)

	Douglas fir-hemlock community.	Oak community
Total weight or volume of ...	About one half that of oak community.	About twice that of fir-hemlock community.
Leaf eating excrement formers (*Lepidoptera* Larvae, *Hemiptera*, *Homoptera*, etc.).	Very few. Aphids only ones ever common.	Many individuals and species.
Humus formers:-Millipedes and centipedes.	Many-Approx. five times number in oak community.	Fairly numerous.
Collembole	Common. Probably about as numerous as in oak community.	Common. Probably about as numerous as in Douglas fir-hemlock community.
Enchytraeidae	Common. Probably about as numerous as in oak community.	Common. Probably about as numerous as in Douglas fir-hemlock community.
Earthworms (*Lumbricidae*)	Common and large.	Rare and small.
Log fauna	Possibly slightly more numerous than in oak community.	Possibly slightly less numerous than in oak community.
Birds.		
Excrement, bodies etc.	20 individuals active per hectare, approx.	35 individuals active per hectare, approx.
Mammals.—Excrement, dead bodies, etc.		
A. Small....................	Common.	More numerous than in Douglas fir-hemlock community.
B. Large....................	Present-wide ranging.	Extripated.

TABLE 12 *(Concluded)*

Animals with weathering reactions.
(after Dirks Edmunds)

	Douglas fir-hemlock community.	Oak community.
Earthworms. *Lumbricidae.*	Common, large (adults 20 cms. or more in length usually at least 50 cm. below soil surface in summer.	Rare, small.
Ants.	Rare. Carpenter ant only species found; it frequents wood.	Abundant, often 50 to 100 per sq. meter. Penetrate deeply into soil.
Collembola.	Numerous (36% of invertebrate population in ground layer).	Numerous about 20% of invertebrate population in ground layer.
Mites.	Fairly common.	Common.

After the worms and other small animals have acted upon the forest litter bacterial decompsition sets in. As shown in Table 14 cast of worms almost always have higher amounts of organic matter and exchangeable bases than the soil in which worms live. This increases fertility of soils that have largest number of worms.

Bacteria derive their energy from carbon content of the litter and nitrogen which is so vital to them forms raw material for the proteins. In fresh litter carbon is usually in abundance so there is always a competition between the various organisms for the available nitrogen.

TABLE 14

Properties of earthworm casts and of soil from forested areas.

(After Lunt and Jacobson, 1944)

Profiles		Total N %	pH	Loss on ignition %	Organic C %	C/N	Available P %	Exchangeable Ca p.p.m.	Exchangeable Mg. p.p.m.	Total Ca %	Base capacity per 100 gms.	Moisture equivalent.
I	Casts.	0.630	5.41	27.6	14.9	23.8	27.4	4280	328	1.00	30.4	48.4
	A_1 Soil.	0.382	4.75	13.6	6.5	17.1	22.3	1183	109	0.94	15.7	27.4
	A_3 Soil.	0.133	4.48	5.8	2.0	14.7	7.8	95	27	0.62	8.1	19.9
	B_1 Soil.	0.086	4.60	4.8	1.3	15.7	9.4	111	26	0.59	7.3	19.0
II	Casts.	0.630	5.35	32.4	17.4	27.6	19.4	5300	511	1.05	34.3	61.2
	A_1 Soil.	0,292	4.65	11.7	5.3	18.0	9.1	844	153	0.58	14.4	31.2
	A_3 Soil.	0.106	4.55	3.6	1.8	17.1	5.1	151	24	0.51	6.3	18.5
	B_1 Soil.	0.039	4.69	2.9	0.6	15.9	3.9	200	69	0.47	5,0	15.4
III	Casts.	0.717	5.00	30.2	16.6	23.1	21.3	3272	354	1.40	29.6	49.6
	A_1 Soil.	0.320	4.43	12.6	6.8	21.1	20.9	224	69	1.46	16.2	26.8
	A_3 Soil.	0.151	4.71	5.6	2.7	17.5	6.8	51	15	1.36	8.2	18.1
	B_1 Soil.	0.07	4.82	3.0	1.0	14.5	13.2	46	12	1.21	3.9	12.6
IV	Casts.	0.523	5.29	25.7	13.4	25.7	16.1	2900	480	0.78	27.5	52.2
	A_1 Soil.	0.314	4.66	11.8	5.0	15.8	7.7	738	227	0.42	13.9	35.1
	A_3 Soil.	0.131	4.69	5.3	2.1	16.0	3.6	323	105	0.48	7.5	24.4
	B_1 Soil.	0.062	4.73	3.4	0.9	13.7	3.3	325	127	0.45	5.9	20.1

The plant carbohydrates are usually found in the form of cellulose and lignin, the decomposition of which is rather slow in strongly acidic soils. It is not yet quite clearly understood how this decomposition goes on in forest soils; there is, however, some evidence available from rotting of timber in forest showing that cellulose and lignin are attacked by a range of *Basidiomycetes,* which may also be responsible for the decomposition of the forest litter. The hyphae of the fungi penetrate and grow into wood thereby breaking it and bringing about decay. Different groups of fungi are involved and may bring about white rot or brown rot. The white rot fungi break up both lignin and cellulose, leaving a whitish remnant of wood composed mainly of cellulose residue and other substances that are not decomposed further. The brown rot fungi leave a brown residue composed of some original lignin of wood, which easily breaks down to powder.

Some experiments conducted in Sweden have shown that *Basidiomycetes* can bring about decomposition of cellulose and lignin of the forest litter. This fact may lead to the presumption that in most temperate regions forest litter is broken down by *Basidiomycetes.*

Most conifer trees possess mycorrhizae which Melin has shown to have common *Basidiomycetes.* Thus, it may seem that in strongly acidic northern conifer forests which normally lack in worms and other litter decomposing fauna the disintegration of organic matter is probably brought about by a range of *Basidiomycetes,* which are aerobic, and here the process of disintegration is analogous to oxidation.

Recent research has shown that mycorrhiza develops in those soils, which normally have a deficiency of nitrogen, phosphorous, potassium or calcium. Some Swedish workers have shown that the development of mycorrhiza in trees is governed by the amount of surplus carbohydrates present in the roots. In rich soils, where nitrogen and phosphorus are abundant, carbohydrates are used in the growth of roots and the growth of fungi is retarded. But if smaller amounts of nitrogen and phosphorus are present, the growth of the plant is not vigorous, and the accumulation of carbohydrates results consequently in the development of mycorrhizal roots.

Rayner and Neilson Jones (1944) have shown "that failure of pines on certain English heathy soils is caused by inhibition of mycorrhizal development, resulting from low concentration and presence of toxic substances of biological origin."

The Indian forest soils are not acidic and the worms are active only during the rainy season. The decomposition of forest litter is probably brought about by the *Basidiomycetes*. Saksena (1955) has worked out the population of bacteria and fungi in the forest soils of Madhya Pradesh, whose data have already been given.

During the transformation of the litter numerous intermediate products are formed and when the fungi are involved numerous fungus hyphae develop which absorb large proportions of nitrogen and mineral constituents in addition to other substances. These constituents are released on the death of fungal hyphae and are added to the humus.

In the decomposition of the littter by white rot fungi the end products are usually carbon dioxide and water. Nitrogen is liberated as NH_4—ion. All these products are easily assimilated by higher plants.

When brown rot fungi are the active agents in the decomposition of the forest litter lignin is formed to which combines nitrogen-containing compounds, forming ligno-protein-complex, as is shown by Waksman. Mattsson however, considers that as lignin combines in addition to proteins with NH_4—ion, the substances formed should be called nitrogenous-lignin-complexes. These products are very difficult to break up and in north conifer forests they form an important part of the humus substances.

The nitrogenous-lignin-complex has acidic properties and its presence makes the conifer forest soils of higher latitudes highly acidic. On account of the high buffering capacity of the humus acids contained in these substances low pH value of the soils is maintained.

The accumulation of raw humus and peat formation occurs in all humid countries at high altitudes. The accumulation of peat has been noticed also on the Himalayas and at the tops of the Nilgiris in the Peninsular India. The causes for this are all well known in European countries and as they may be of interest to ecologists in India a short account may profitably be included here.

Discussing the soil forming processes on British mountains, Pearsall (1954) attributes peat accumulation to three main environmental factors, namely, water-logging, leaching and low temperatures. The effect of these three different factors is generally similar in reducing the activity of the soil micro—organisms, which affect the vegetation and in turn animal life.

The soil flora mainly consists of bacteria and fungi, which bring about decay of remains of dead plants, or animals. In forest soils larvae of flies and earthworms initiate break-down of plant material which is later acted upon by fungi bacteria and protozoa which by browsing upon fungi and keeping them in check help in the transformation of plant material. Several insects as springtails and myriapods (shown by Romell in N. America), etc., play their part in the natural economy of the soil.

This population of the soil is aerobic in character and requires oxygen from the air for breathing. In a normal soil, therefore, the products of transformation of plant material usually must contain oxygen and produce CO_2, which escapes to the air. Nitrates, sulphates, phosphates, etc., remain in the soil. Of these substances nitrates on account of their greatest requirement by plants for growth determine whether a soil is fertile or not. The fertility of the soil is thus an index of nitrate production capacity of soils. Although, plants can absorb a certain amount of ammonia, which is the first product in the breakdown of nitrogenous organic material it seldom remains stable in soils, and it is quickly converted into nitrates by the action of the nitrifying bacteria. The nitrogen fixation can also be brought about in soils by the nodule bacteria of leguminous plants.

The harmful effects of waterlogging, leaching and low temperatures on the decreased soil fertility and poor vegetal growth are similar. Waterlogging mainly reduces the amount of oxygen from the medium, as a result of which aerobic soil population is reduced to the minimum. In such conditions a few anaerobic organisms, which can maintain themselves in the absence of oxygen, occur. The products of decomposition of the organic matter of the anaerobic organisms do not contain oxygen; but usually methane (CH_4), ammonia (NH_3), sulphuretted hydrogen (H_2S) and sometimes phosphine (PH_3); substances without oxygen, are formed. These substances are not utilised by most higher plants and are toxic to them.

Waterlogged and aerated soils thus represent two different biological systems: one is reducing and the other is oxidising and the level of oxygen present in these determines the type of vegetation that will occur on them.

It may be useful to represent these two systems in a graphical manner below:—

Oxidising soils contain		*Reducing soils contain*
Nitrate.	...	Ammonia.
Carbon dioxide.	...	Methane (marsh gas).
Sulphate.	...	Sulphide.
Phosphate.	...	Phosphine.
Ferric iron.	...	Ferrous iron.

It may be realised that during the rainy season in India many of the normal oxidising soils may become reducing at times and oxidising at others, when, the rains stop and evaporation sets in. The die-back of tree seedlings during the months of August and September may probably be investigated in the light of biological processes sketched above.

7

Organic matter system of the soil

Soil processes and plant nutrition are intimately related with the organic matter or humus content of the soil. Plants and animals of all groups living outside and within the soil add organic matter to it, upon their death and decomposition. The biological system of the soil as shown in the preceding chapter, is ever so busy in breaking down litter, roots, remains of worms, insects, protozoa, bacteria, fungi and other organisms and mixing up the residues with the mineral particles. This breaking down process, or decomposition in which both physical and chemical processes are involved is known as humification. The finer particles of the residues form the humus, which is incorporated in the soil.

(1). **Humus**. Humus is a complex aggregate of brown to dark coloured organic particles of different sizes, arising from decomposition of plant and animal remains, which have lost structural and cellular organisation. Some authors, however, include the stages of decomposition also in it.

Humus is a storehouse of important plant nutrients, and is also capable of releasing these from the organic complexes of the soil. Its main attributes are colloidal, buffering properties, combining power with bases (adsorption), redox potential, ability to supply trace elements, improving soil structure and moisture holding capacity, etc.

(2). **Humification.** In bulk the vegetable matter is far in excess of animal matter. It is only through the former that the solar energy is fixed and kept in circulation among all the organisms of nature. Hence the major sources of humus are forests, grasslands, other low vegetation and aquatic plants.

(a) *Forest litter.* In most deciduous forests litter is available anually as a crop but in evergreen forests leaf fall is gradual and is not as pronounced as in the case of the former type of forests. In addition to leaves, small and large twigs, parts of fruits and seeds also contribute to some extent to the soil organic matter.

The amount of litter production in a forest. The extent and the quantity of litter produced by a forest depend upon the species, their phenological habits, the age of the stand and the number of trees occurring per unit area. In a general way, it is dependent on almost all the factors of the locality. An interesting idea of organic matter production by vegetation in different types of forests in the world may be got from Table 15.

(ii) **Quality of the tree litter. Organic**:—The properties of organic matter and its biological activities largely depend upon the chemical constituents present in the tree leaves. Most plant parts consist of a great amount of carbohydrates and lignin. The carbohydrates are mainly polysaccharides which are built up of simpler carbohydrate molecules, such as sugar and are linked up into larger and more complicated molecules. Of these complicated molecules of carbohydrates, cellulose is the most important, though hemicelluloses and pentosans also play an important part in the formation of humus. The carbohydrates as well as lignin consist merely of atoms of carbon, hydrogen and oxygen. In the tree litter these are always mixed with proteins which are composed, in addition to, carbon, hydrogen and oxygen, of nitrogen and small amounts of sulphur, and in some cases of phosphoric acid.

In addition to these substances, fats, resins, waxes and organic acids may also be present in tree litter.

The chemical composition of some typical European forest residues is given in Table 16.

TABLE 15

Organic matter production by vegetation (Roots not included unless stated).

(Partly from Jenny).

Type of vegetation	Annual production tons per acre	Author.	Remarks.
Oakwoods. ...	2.12-12.45	Pearsall (1945)	Hertfordshire U.K. dry wt.
Average forest (leaves, wood, twigs.)	267	Henry	German forests air-dry.
Beech forest — wood. ...	1.42	Ebermayer.	Central Europe, air-dry.
Beech forest — leaves. ...	1.47	,,	
Pine forest — wood. ...	1.42	,,	,,
Pine forest — leaves. ...	1.42		
White pine needles.	2.09	Lunt.	Continent, 1933, oven-dry.
Beech-birch-aspen forest. ...	2.85	Henry.	Leaf litter only Germany, air-dry.
Tropical primeval forest (leaves, trunks, roots). ...	11.1	Hardon.	Java, estimate, moist.
Monsoon forest. ...	22.2	Vageler.	Estimate, moist.
Tropical rain forest.	45.90	,,	,, ,,
Tropical rain forest.	100	Champion (1936).	Estimate.
Typical tropical deciduous forest.	20	,,	,,
Tropical forest on red soils. ...	2½	,,	,,
Sal forests of Dehra Dun. ...	2.4	Puri (1954).	Leaf litter, actual determination in quadrats at leaf-fall, air-dry.

TABLE 16

Approximate chemical composition of some typical forest residues on percentage of dry material.

(Adapted from Waksman, Tenny and Stevens, 1928).

Chemical constituents.	Pine needles (green).	Oak leaves (green).	Oak leaves (mature).	*Hypnum* moss.	Litter of ever-green forest.	Oak leaves, decomposed 6 months in lab.	Humus layer of the forest soil.
Ether-soluble portion.	7.65	7.75	4.01	4.58	6.38	5.34	3.04
Cold & hot water soluble organic matter.	13.02	22.02	15.32	8.41	4.06	7	13.84
Hemicelluloses.	14.68	12.50	15.60	18.92	18.23	18.52	8.71
Celluloses.	18.26	15.92	17.18	24.75	15.72	11.44	8
Lignins.	27.63	20.67	29.66	21.13	38.38	47.64	54.21
Porteins.	8.53	9.18	3.47	4.16	5.79	10.66	12.33
Ash.	3.08	6.40	4.63	4.33	2.68	—	—

During the decomposition of the litter, water soluble substances are first attacked by the soil population and the proportion of these and celluloses decreases. Proportion of lignin and protein, however, increases.

The increased proportion of lignin and proteins in the humus layer given above may also be due to the climatic conditions and the scarcity of the right type of soil population.

Inorganic (Mineral) :—In addition to the organic compounds described above, tree litter contains a proportion of inorganic substances such as lime, potash, magnesia, iron, manganese, silica, copper, aluminium and small amounts of practically all the elements, which the living organisms require for their growth. These are left behind when plant material is ashed, and hence they are called the ash constituents.

Of the inorganic substances, calcium forms by ar the greatest amount of the ash in a large majority of plants.

The inorganic constituents in the litter of a few Swedish tree species are given in Table 17.

TABLE 17.

Composition of some important forest litter in Swedish forests, on percentage of the dried material.

(From Tamm. 1950).

	Pine needles	Spruce needles	Birch leaves	Aspen leaves	Alder leaves
Ash	1.63	8.40	5.58	9.53	4.96
Iron oxide (Fe_2O_3)	0.024	0.030	0.017	0.015	0.074
Alumina (Al_2O_3)	0.182	0.046	0.020	0.044	0.140
Mangano-oxide (MnO)	n.d.	0.21	0.072	0.038	n.d.
Lime (CaO)	0.821	2.01	1.99	3.34	2.35
Magnesia (MgO)	0.082	0.218	0.56	0.44	0.539
Potash (K_2O)	0.083	0.455	0.61	0.98	0.398
Soda (Na_2O)	0.054	0.027	0.013	0.012	0.068
Copper oxide (CuO)	n.d.	0.001	n.d.	0.002	n.d.
Silica (SiO_2)	0.232	3.29	0.373	1.56	0.99
Phosphoric acid (P_2O_5)	0.083	0.158	0.143	0.129	0.188
Sulphur (SO_3)	0.006	0.42	0.23	0.25	0.20
Chloride (Cl)	n.d.	0.07	0.012	0.16	0.024
Boron (B)	n.d.	n.d.	n.d.	0.002	n.d.
Nitrogen (N)	0.46	0.96	1.18	0.58	2.38

From the above table, it is apparent that calcium and nitrogen are mostly in high proportion in tree litter. In the development of soil humus as these two constituents play the most conspicuous part, the percentage of these alone may give a good indication of the humus conditions and fertility level in a forest. This was shown by Puri (1951) study of some in his important tree species of temperate Himalayas.

The amount of ash, calcium, nitrogen and C/N ratio of litter of species from England, North Western Himalayas and of mature leaves collected from Darjeeling, Dun

Valley, Bombay and Madras State forests are given in Table 18.

TABLE 18

Composition of plant litter ; on % dry material.

(From Puri, mainly unpublished data).

Species	Ash	Carbon	CaO	Nitrogen	C : N	Region
I—Litter.						
1. *Tilia cordifolia.*	10.45	47.2	4.38	2.17	21.1	England
2. *Fraxinus excelsior.*	8.43	48.3	3.38	1.67	22.2	,,
3. *Pinus sylvestris.*	2.91	51.3	0.78	1.24	41.2	,,
4. *Pinus longifolia.*	3.57	48	0.56	1.54	31	N. W. Himalayas
5. *Quercus ilex.*	3.85	49	1.9	1.19	41	,,
6. *Abies webbiana.*	4.70	42	1.59	1.56	26	,,
II—Leaves.						
7. *Elaeocarpus sikkimensis.*	4.51	48	0.94	2.38	20	Darjeeling
8. *Michelia cathcartii.*	6.68	47	1.18	2.34	20	,,
9. *Cryptomeria japonica.*	6.19	48	2.64	0.89	54	Cultivated
10. *Acacia catechu.*	5.95	46.08	2.03	4.18	11.02	Dun Valley
11. *Albizzia procera.*	5.54	46.79	1.55	3.21	14.67	,,
12. *Bassia latifolia.*	6.29	47.08	1.93	1.89	24.9	,,
13. *Anacardium occidentale.*	2.7	49	0.49	1.45	35	Bombay
14. *Lagerstroemia parviflora.*	6.24	48	1.98	0.81	59	,,
15. *Tectona grandis.*	8.95	46	2.02	1.01	46	Cultivated Ghera
16. *Albizzia odoratissima.*	6.93	46	1.29	.41	19	Sinhgadh
17. *Semecarpus anacardium.*	9.33	46	1.89	1.47	31	,,
18. *Bridelia retusa.*	6.5	47	2.41	1.63	29	,,

Similar studies on a large number of trees show that there is a good deal of difference among the various species in the same type of forests in both calcium and nitrogen contents. This may account for the different kinds of humus occurring in different forest types, although the effects of climate and soil population are also important in considering the type of humus.

(iii) **Circulation of mineral nutrients in the soil:**

It is apparent that the mineral ash matter of the litter is derived from the soil itself and on humification is redistributed in the surface layer. Thus, a large amount of nutrients is kept in circulation and is readily available to the plants again. In Table 19 are given the amounts of nutrients that are added to the forest floor in different types of forests.

TABLE 19

Amounts of litter and some of their components, added to one acre of the soil in the course of a year/lbs. per acre.
(after Tamm, 1950).

Origin	Litter dried	Nitrogen	Lime CaO	Potash K_2O	Phosphorus P_2O
Spruce stand, 40 years old near Oslo, South Norway (Mork)	2920.5	34.3	43.1	10.3	5.6
The same spruce stand another year (Mork).	2981.0	35.0	44.6	9.6	6.9
Spruce, stand, 60 years old, Veldre Norway (Mork)	1680.2	19.3	40.5	4.6	3.0
Alpine spruce stand, 140 years old Hirkjolen, Norway, Averages from 6 years (Mork).	1307.3	11.8	21.7	2.9	2.3
Alpine pine stand, 200 years old Hirkjolen, Norway, Averages from 6 years (Mork).	714.6	11.2	10.3	1.10	2.34
Birch stand 35 years old near Stockholm, Sweden (Hanson and Knudsen).	1487.3	19.6	29.7	9.3	2.2
Spruce stand 90 years old, near Stockholm, Sweden (Norming & Lindberg).	2756.0	26.5	56.8	12.5	4.4
Aspen stand, near Stockholm, Sweden (Anderson and Enander).	1614.9	9.4	53.8	15.8	2.0

It may be interesting to compare the approximate amount of nutrients added to the forest soil per acre in the form of leaf litter during a year with the quantities of fertilizers used in agricultural practices in Sweden. It is estimated that 40 lbs. of nitrogen correspond to approximately 260 lbs. of calcium nitrate; 50 lbs. of CaO correspond to about 100 lbs. of ground limestone; 10 lbs. of K_2O to about 25 lbs. of fertilizer; and 5 lbs. of P_2O_5 to 25 lbs. of 20% superphosphate.

It is apparent that plants are able to extract tremendous quantities of fertilizers from the lower layers of the soil and bring them into biological circulation. From the lists of forest trees given in the preceding pages it seems clear that on the whole conifer species on account of their containing low amounts of lime have a less favourable effect on soil humus than most broad leaved species. Among the broad leaved species the majority of evergreen and semi-evergreen types contain lower amounts of calcium than the pronounced deciduous types. It may thus be possible to classify species in small areas into groups with regard to their effect on the humus content of soil. For instance, those species which have high amount of foliar Ca may be called exacting species and they produce highly fertile humus layer, while those having low amounts of Ca in their leaves may be called non-exacting. The humus conditions in such forests are not very active. These observations from the northern temperate forests of Europe and America may not strictly apply to conditions in the tropical forests of this country, where, although, species have low amounts of foliar calcium, vegetal activity in the forest floor does not by any means seem to be less active. Such studies in this country will be of great importance to forestry.

(*b*) **Plant and soil calcium and nitrogen.**—The amount of foliar Ca is probably a good index of the Ca—status of soils. It has been shown that exacting or calcicole species that grow on soils with higher amounts of soil Ca contain higher amounts of foliar Ca. The Ca content in most of the calcicolous plants studied by Puri (1950) in deciduous woodlands is about 1.5% ; soil pH in the habitat of these plants is mostly above 7. Soil pH and Ca content in leaves of some calcifugous plants in the same forests is low. These plants have mostly foliar Ca below 1% and a soil pH between 3.3—5.3.

In Table 20 are given mineral composition of a few weeds and crop plants when grown on acidic and basic soils.

TABLE 20

Mineral composition of some species on dry weight basis as %

Species	N	CaO	K_2O	S	P_2O_5	Investigators.
—Weeds.						
1. *Argemone mexicana* (on base poor soil).	1.01	1.89	1.33	0.53	1.36	Singh and Singh, 1950
2. *Cassia occidentalis* (on lime rich soil).	3.08	5.65	2.31	0.54	1.56	,,
3. *Digera arvensis* (on lime rich soil)	3.24	4.25	3.15	0.55	1.63	,,
4. *Lindenbergia polyantha* (on lime rich soil).	—	4.23	—	—	—	Misra and Siva Rao, 1948
II—Crop plants.						
5. *Sorghum vulgare saccharatium.*	0.96	0.41	—	—	—	Parker and Truog, 1920
6. *Zea mays.*	1.39	0.60	—	—	—	,,
7. *Triticum vulgare.*	2.11	0.34	—	—	—	,,
8. *Beta vulgaris.*	2.23	0.65	—	—	—	,,
9. Barley on acid soil.	2.88	0.35	—	—	—	Bender and Eiso-umonger. 1941
10. Barley on basic soil.	3.16	1.32	—	—	—	,,
11. Wheat on acid soil.	2.83	0.44	—	—	—	,,
Wheat on basic soil.	3.34	0.80	—	—	—	,,

TABLE 21

Mineral content of plants on dry weight basis.

Species	Soil pH.	Plant part.	Ash.	Al.	Fe.	Mn.	P.	Ca.	K.	Investigator.
I. Calcicolous										
1. *Medicago sativa.*	5.9	leaves	9.19	0.013	0.015	0.005	0.27	1.80	1.61	Hsich-yu Hon and Merkle, 1950
2. *Dispercus sylvestris.*	8.0	,,	12.12	0.014	0.019	0.005	0.54	2.09	2.74	,, ,,
I. Calcifugous										
3. *Lycopodium clavatum.*	4.1	whole	4.25	0.263	0.028	0.032	0.14	0.12	1.20	,, ,,
4. *Rhododendron maximum.*	4.2	leaves	4.45	0.019	0.015	0.123	0.09	1.09	0.39	,, ,,
III. Indifferent to calcium.										
5. *Sambucus canadensis.*	4.0	,,	8.59	0.055	0.076	0.038	0.11	1.59	1.23	,, ,,
Sambucus canadensis.	6.3	,,	10.03	0.039	0.029	0.016	0.34	1.62	1.88	,, ,,
6. *Zea mays.*	3.7	,,	6.17	0.028	0.023	0.119	0.23	0.50	1.79	,, ,,
Zea mays.	8.1	,,	6.09	0.026	0.024	0.012	0.27	1.05	0.47	,, ,,

In all the cases plants grown on basic soils have a higher amount of calcium. This is also found true for nitrogen. Some data given by Hsich-yu Hon and Merkle (1950) in Table 21 show similar results. Plants may be classified on their calcium relationship into calcicoles, calcifuge and indifferent and soil calcium. Very little is known about the third category of plants.

(3) **Grassland and other soils :** In grassland soils it is obvious that the quality and quantity of the vegetation will set up a more superficial circulation of mineral nutrients on account of humification. The shoots of the grasses are generally grazed hence it is mostly the root matter which will supply the organic material. Data of analysis for two grasses are given in Table 22.

TABLE 22

Mineral composition of two common grasses on % dry weight basis.

Pandeya (1954).

Species	Ash	SiO_2	Fe_2O_3	CaO	Al_2O_3	N
1. *Bothri ochloa pertusa.*	10.5	3.26	1.63	1.01	2.17	4.15
2. *Dichanthium annulatum.*	11.5	4.4	1.70	1.54	1.85	5.12

(4) **Aquatic plants :** The chemical composition of a few aquatic plants are given in Table 23.

TABLE 23

Mineral composition of some aquati lants on % of dry weight basis.

Misra (1938)

Species	N	Ash	SiO_2	CaO	MgO	Fe_2O_3	Mn_3O_4	Al_2O_3	PO_4
Chara fragilis.	8.15	18.87	4.38	[illegible]	[illegible]34	1.38	0.26	0.92	0.82
Myriophyllum sp.	5.99	21.19	11.48	3.15	0.11	0.61	0.11	1.36	0.71
Potamogeton perfoliatus.	7.57	16.82	1.95	1.91	0.42	0.44	0.03	2.19	11.4
Nymphaea alba.	1.90	11.20	1.60	1.04	0.26	0.04	Tr.	1.10	...

It has been shown in these cases that the mineral salts are absorbed from the mud which is the rooting medium. There exists a complex system of mud-plant and water through which the nutrients are kept circulating. Exchanges between mud and water have been studied by a number of workers in England.

(*c*) **Carbon : Nitrogen ratio.** The carbon nitrogen ratio in forest litter is on the average about 30. In case of Leguminous plants it is no doubt lower. However, the C/N ratio in case of humus has been shown by a number of workers (cf. Waksman) to range between **8-14.** Hence, it is evident that in the process of humification proportionally larger amounts of carbon are lost and nitrogen is conserved. In fact, carbon is largely used up in 'soil respiration' by the micro-organisms which consume decomposing organic matter as food. The same ratio holds good for grasses and herbs also. In all these cases humification takes place under aerobic conditions.

The aquatic plants are comparatively richer in nitrogen and the body C/N ratio in these is not far off from the C/N ratio of the mud humus. It is, therefore, evident that under sub-aqueous anaerobic conditions humification proceeds in a manner that carbon and nitrogen losses are balanced. As has been already observed carbon may be lost in the form of marsh gas and nitrogen in the form of ammonia. Organic matter from terrestrial sources accumulating under water gives rise to peat.

(*d*) **Mull and Mor.** The biological and chemical status of the humus of the forest floor has been of great interest to ecologists and hence a brief account of the subject as developed in Europe is given here.

Mull (Danish spelling Muld), and *Mor* (Danish Mor, or Maur, also Mear) are tw[illegible] classes as accepted by the International Union of [illegible]est Research Organisations in 1933, and are easily distinguished in the field by their characteristic structural differences and biological potentialities as reflected in the types of vegetation they bear. Mull is typically "a porous, loose, friable mass with a crumby structure, and consists of an intimate mixture of humus with mineral soil. The organic matter content of the mixture is comparatively low (in Denmark, below 10%). The deposit is inhabited by large earthworms, such as *Lumbricus terrestris*, and supports a rich vegetation of geophytes."

"The Mor comprises matted humus layer usually very low in inorganic matter and lying like a carpet on top of the mineral soil. The deposit is more or less tough or compact, or both, not porous and friable like the Mull. The large earthworm species are missing. The vegetation is strikingly different from that on the Mull type, not only in the absence of the typical Mull plants but also in the occurrence, often even abundance, of plants (Ericaceae, Mosses, etc.) absent or rare on the Mull" (Romell and Heiberg, *loc. cit.*) Mor is also called *raw humus*. The term duff was used for Mor by Romell and Heiberg but later Romell agreed to the adoption of the word Mor in Muller's sense.

In waterlogged swampy soil at higher altitudes in northern countries a third type of humus, called *peat* is distinguished. Peat is variously referred by different authors as peaty soil or organic soil.

A large number of intermediate forms between the Mull and Mor have been recognised in different forest types. Thus, mull-like-mor is a humus which in the surface may look like mor but the lower layer has properties of mull.

These types of humus layers have been studied in detail in north European countries, especially Denmark and Sweden, and in North America. Some work was recently conducted by Puri (1950) in English woodlands. A brief summary that will be presented here is based on the works of Muller, Hesselman, Tamm, Romell, Heiberg, Pearsall and Puri. Unfortunately, no work of this type has so far been done in India, except that some preliminary studies were published by Suri and some brief studies in continuation of Suri's work have been published more recently by Puri and Gupta (1951). Further work in this direction will be of undoubted value.

Mull and Mull like Mor

This type of humus layer is usually found in broad-leaved deciduous forests of temperate regions. Usually, they occur in the forests developed in plains. However, in some high level forests in Sweden, Mull-like-Mor type of humus layer may be found. It is associated with a brown earth profile, though some podsol profiles may also have a Mull type of humus layer. In northern regions where the humus type is predominantly Mor, small oasis of Mull may be present locally in lime-rich situations, flushed areas or in depressions with a deep loamy soil.

In India, although, no work comparable to that of Swedish and Danish ecologists has been done, it appears that in most of the Indian forest types the humus is comparable either to Mull or Mull-like-Mor. Mor may be found in high level conifer forests, but it is doubtful whether Mor from these areas will be as inactive as that found in many high level conifer forests of northern countries (Puri and Gupta, 1951).

Mull layer is characterised by its granular structure and ecologically it is a richer type than the Mor. As already noted, Mull has a great abundance of animal life, especially worms and bacteria. A certain number of white-rot-type of Basidiomycetes may also be present. These organisms are able to bring about a complete transformation of tree litter producing a dark coloured humus which is rich in nitrogen as nitrates and kations, especially calcium which are adsorbed on the colloidal complex. The reaction of Mull type is usually less acidic or neutral.

Mull is less acid and sometimes alkaline with comparatively low amount of humus content and high amount of nitrification. The upper and lower layers in Mull are not distinct.

Chemical properties of some Mull and transition-to-Mor soil with the characteristic ground flora communities and characteristic tree seedling growth for some English woodlands is given in Table 24.

Relationship between pH value of the soil with the distribution of ground flora communities and tree seedlings is given in figures 20. The pH Curves, as expected, show a considerable overlapping between different ground flora communities, as well as tree seedlings growth, indicating that soil pH is not the only factor that determines the distributional limits of the various plant communities. The data, however, show that pH values of Mull and transition to-Mor-soils are different.

In the attempt to discover the causes for differences in acidity and other humus properties in Mull and Mor soils found in English woodlands detailed analysis of various factors

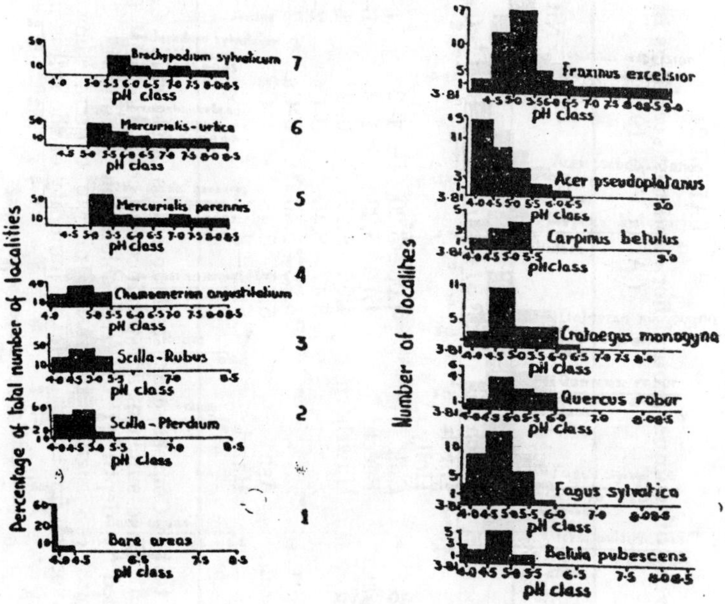

Fig. 20

TABLE 24

Chemical properties of Humus, ground flora and tree seedlings in some English woodlands.
(After Puri, 1950)

Humus type	Ground flora community	Tree seedlings	pH	Thio-cyanate	Nitrate	Organic matter	Summer relative humidity
Transition to Mor	Bare areas	Beech Birch Holly	3.81—4.01	1–3	0–1	14.3±2.3	2.8±0.2
Mull	*Scilla-Rubus*	Oak	4.08—5.20	1–2	1–3	9.7±2.6	3.6±0.8
	Scilla-Pteridium	Beech	4.14—5.23	1–2	1–2	8.7±2.1	3.7±1.4
	Chamanerion	Hawthorn	4.08—5.27	0–2	1–3	7.1±2.4	4.7±2.2
True Mull	*Urtica*		5.47—8.45	0–Ca	2–4	6.4±3.2	5.0±1.5
	Brachypodium		5.28—8.40	0–Ca	2–4	5.3±2.6	5.4±2.0
	Urtica-Mercurialis	Ash	5.29—8.45	1–Ca	1–4	5.8±2.0	5.7±2.1
	Mercurialis	Hazel	5.25—8.45	1–Ca	1–4	4.8±1.8	6.1±2.6

NOTE :—Thiocyanate test of Comber measures base states of the soil. *Ca*–calcareous soil giving effervescence with dil Hcl. *O*, is lime deficient, Base–sufficient soil produces no coloration with Comber's reagent. *1-3* signify increasing base deficiency as indicated by depth of the red coloration produced. Nitrate is tested by diphenylamine test ; *O*–means no coloration. 1–3 signify increasing nitrate content shown by depth of the blue coloration produced. Organic matter is loss on ignition. Relative humidity of the soil is moisture content/loss on ignition.

were undertaken. A good correlation between different properties of the humus with topographical features as shown in Figure 21 was obtained and this was correlated with the distribution of the various ground flora communities and tree seedlings (Figure 22). On the whole, it appears that the transition-to-Mor soils occur at higher levels and Mull are found at lower

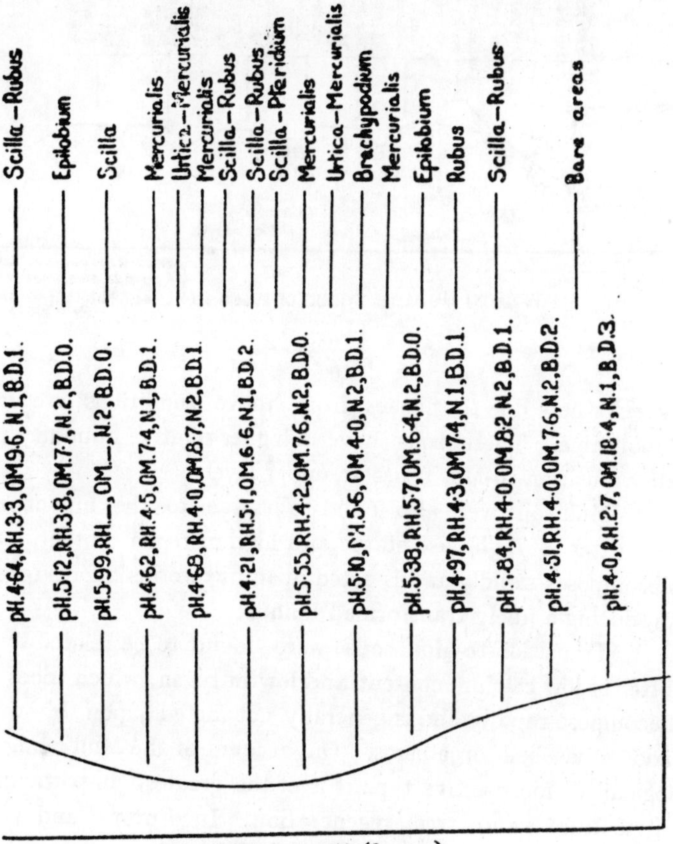

RH.— Relative humidity (Summer)
OM.— Loss on ignition (Organic matter)
N. — Nitrate content
B.D.1–3 — Increasing base deficiency
B.D.0 — Base saturated

RELATIONSHIP BETWEEN HUMUS CONDITIONS & GROUND FLORA COMMUNITIES. (After Puri, 1950.)

Fig. 21

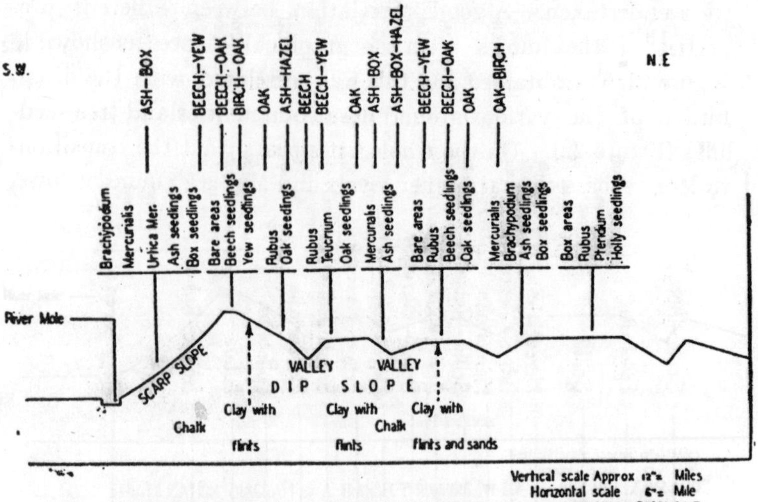

FIG. STRATIFICATION DIAGRAM OF A TRANSECT RUNNING SW-N.E. ON BOX HILL, ENGLAND. (After Puri)

Fig. 22

levels, where due to flushes from above conditions are more favourable. These soils have a higher relative humidity than the transition-to-Mor types at upper levels.

Mull soils were also found to be due to the influence of tree litter of high base status and high nitrogen content, which decomposes rapidly and is acted upon by worms more readily, producing a finely transformed humus.

Transition-to-Mor soils were found to be related to tree litter of low calcium content and low nitrogen, which does not decompose rapidly and is generally not acted upon by worms and other soil organisms. On account of the Mull being ecologically a more active type it is of the greatest importance in silviculture and tree regeneration. In Europe and North America most of the broad leaved deciduous tree species regenerate themselves abundantly on Mull, or Mull-like-Mor soils. These types of humus have the capacity of self seeding in small gaps in the canopy, thereby showing that these forests are eminently suited to the selection systems of management and silviculture. Clearfelling in Mull types has been shown by Swedish workers to create conditions unfavourable for the establishment of tree species; as a luxuriant growth of ground flora species results. On account of the humus being intimately mingled with the mineral soil, Mull creates the most favourable seed bed for the germination of tree species. Good germination of most of the important broad leaved tree species is reported in temperate regions on Mull soils.

Silvicultural and management practices alter the conditions in Mull, but in the course of time original conditions are restored.

Most conifers are not able to grow in Mull soils on account of their being too rich in mineral food.

Biologically, Mull constitutes a distinct type, as decomposition of forest litter here is brought about predominantly by bacteria and worms. Some white rot fungi occur in this system, but they are not so active as other organisms. Protozoa keep the growth of fungi in check. Most nitrifying bacteria occur in Mull soils as a result of which nitrification is very active. The ammonia formed is rapidly oxidised to nitrates, which are then easily absorbed by growing plants.

Mor

Mor type of humus layer is generally found in conifer forests of northern latitudes. They are usually associated with podsol profiles. In cold temperate zones where Mor is the predominant type, oasis of Mull may occur in localities of favourable drainage or suitable soil conditions.

Mor is biologically distinct from Mull, having predominance of fungi, chiefly brown rot Basidiomycetes, and Ascomycetes. Some bacteria and other fungi and lower animals may also be present to some extent, but due to the production of humic acids unfavourable to the growth of micro-fauna and flora the population of these after a time is greatly reduced. The decomposition of litter is slow and ligno-nitrogen complexes which are acidic in nature are formed. Generally, Mor soils are comparatively dry, more acidic and normally lack nitrates. However, when activated some nitrates may be formed in Mor soils. Most nitrifying organisms and Leguminous plants are not able to flourish in Mor soils, but heather (*Calluna vulgaris, Vaccinium*) and other Ericaceous plants occur normally in this type of humus. Regeneration of most broad-leaved deciduous tree species does not take place in Mor soils, however, most conifers, birch and *Pyrus aucuparia* regenerate predominantly only in these soils.

In Mor type organic matter is not intimately mingled with the mineral soil, but a layer of unincorporated, mostly undecompsed forest litter, called F layer (Fermentation layer) rests on the top of the soil. The lower one is called H layer.

Humification conditions in Tropical India

Due to higher temperature and moisture supply mainly available during the rainy season, the course of humification in tropical parts of India seems to be different. Leaf fall takes place from February to April when the soil is dry. The dry vegetable matter drifts with the wind, accumulating at sheltered places. Due to mechanical forces it breaks into finer pieces which may be attacked by white ants. During the warm humid rainy season humification starts at once and within two months most of the matter is transformed into fine humus. It is during this time that earthworms are most active. The decomposition is so rapid that the soil organic matter remains ever very low. In the absence of a definite humus layer on the surface of the soil it is doubtful if the terms Mull and Mor can be applied to these soils. At high altitudes on the Himalayas litter and coarse humus may accumulate on the surface temporarily but rapidly decomposed during the summer when insolation and surface temperature attain surprisingly high values—120° F at 7,000 feet as it was shown by Gorrie (1933) in the Bashahar Himalayas.

8

Soil Solution System

Within the mass of soil, there exists a weak solution of various salts in water. The composition, concentration, quantity and the way in which it is held and maintained around the roots and soil particles controls the entire growth of plants in nature. The solvent water is derived from precipitation available on the surface periodically; and its movement and collection along solid masses of rocks and earth is very important. The salts arise as a result of weathering of the soil particles and humification. The availability of both water and salts to the plants depends upon the texture, structure, chemical nature and biological activities of the soil and on a number of chemical and physical forces working within.

WATER

Water economy in plants :

The plant body is constituted largely of water. It may amount from 40-70 per cent in weight in terrestrial plants and upto 95% in aquatic plants. In fact, the plant system may be considered as continuous columns of water held up by cellular structure. It is present in cell sap, protoplasm, cell wall and vessels of the plant. It is not in a static condition but is moving all the time according to tensions and stresses set up by the salt concentration of the solution and the colloids within the tissues. It maintains hydrostatic pressure or turgidity within cells, provides a medium for movement of ions and molecules and for all metabolic reactions.

There is a considerable loss of water from the plant body on account of transpiration and this loss is made up by absorption from the soil. Thus, the soil water is being lost through the plants. The plant and the soil have a continuous water system but between them there is struggle for its absorption and retention. When under these stresses the continuity is broken a deficit water budget in the plant leads to its desiccation and death. An understanding of the water budget of the soil and the plant is of very great importance in ecological studies.

Source of water. The ultimate source of water for plant growth is atmospheric precipitation, part of which is absorbed by the soil and part lost by run off and evaporation. A hydrological cycle is represented in Fig. 23 after Thornthwaite.

Vegetation and rainfall: Vegetation usually intercepts fairly large quantities of rainfall coming on to soil. The crown development of trees influence the intercepting of rainfall. There is no Indian data on the subject, but some foreign data taken from Clark (1940) are given in Table 25.

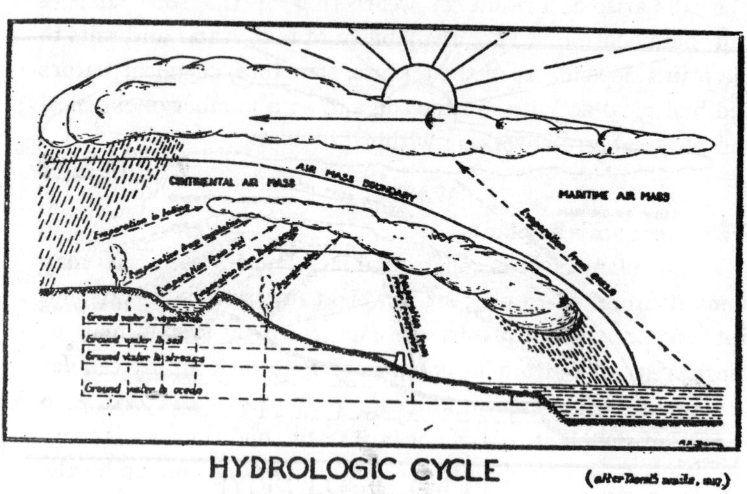

HYDROLOGIC CYCLE

Fig. 23

TABLE 25

Amount of water intercepted by an acre of vegetation under natural rainfall.

(From Clark, 1940).

Date	Vegetation	Rainfall inches	Intercep-tion per cent	Water held tons per acre
July 2	Big blue stem	1.61	68	123.7
July 19	Big blue stem	0.58	73	47.4
June 1	Wheat	0.80	51	45.9
June 22	Wheat	1.48	33	55.1
June 1	Alfalfa	0.84	36	34.2
July 2	Alfalfa	1.80	26	52.8
July 2	Oats	0.74	45	37.6

It will be seen that between 26-73% of the rain falling in the area is intercepted by such low plants as alfalfa, oats, and grasses.

A square meter of big blue stem grass intercepts, still higher percentages of rainfall. In light showers the percentage of interception is usually larger and in very light showers hardly any water reaches the soil surface. In a heavy shower, however, interception is naturally of the smaller order (Table 26.)

The maximum rainfall interception capacity of different ground flora species, cereals and weeds is given in Table 27 on page 135. These data show very clearly the great part vegetation plays in reducing the impact of rain drops on soil and in reducing soil erosion.

Except some mosses and lichens which can absorb moisture from humid air higher plants take their water only after it has been absorbed in the soil. If other conditions are favourable regions of high rainfall have higher amounts of water in the soil, and in humid regions water table is generally high. In arid regions, where the amount of rainfall is low, water table is either very low or there is no water table for most of the year. For the growth of plants, therefore, regions

TABLE 26

Interception of rainfall by a square meter of big blue stem grass under natural rainfall.

(From Clark, 1940).

Date of reading		Inches	Precipitation Character.	Interception %
June	6	0.07	Light shower.	83
,,	11	0.02	Light shower.	90
,,	14	0.45	Hard shower.	70
,,	16	0.06	Light shower.	80
,,	21	0.39	Hard shower.	67
,,	22	1.78	Heavy rain followed by showers. and mist.	68
,,	25	0.14	Several very light showers.	79
,,	26	0.10	Light showers.	87
July	2	1.61	Heavy rain followed by showers.	68
,,	6	0.15	Light showers.	78
,,	19	0.58	Rain and showers.	73
,,	22	0.10	Several very light showers.	76
,,	26	0.12	Very light showers.	80
Aug.	4	0.04	Very light showers.	95
,,	15	1.20	Heavy rain and showers.	63
,,	16	0.18	Light shower.	77

with a high water table are most suitable and support luxuriant vegetation of high trees. In arid regions, plants develop very long root systems, which probe into deeper layers of the soil and their growth is possible only, if the roots are able to reach the water table. *Prosopis spicigera*, *Acacia arabica*, and *Salvadora oleoides* are some of the species the roots of which reach the water table before they get established in arid regions of north-west India.

The amount of water held in the soil depends on the nature of the soil. A large part of the rainfall usually runs off from the surface according to its topography and a part is lost to the atmosphere through evaporation. The water which has

Table 27

Maximum rainfall interception capacity of a square foot of plant materials in grams.

(After Clark, 1940).

A. *Living Plants.*

Plants	No. of single living plants
Prostrate pigweed.	47
Spotted spurge.	52
Knotweed.	71
Puncture vine.	74
Slough grass.	77
Purslane.	107
Foxtail.	137
Bind weed.	137
Prairie dropseed.	138
Burning bush.	148
Tall panic grass.	148
Buffalo grass.	153
White sweet clover.	169
Saltbush.	182
Tall goldenrod.	214
Bi g bue stem.	217
Stink grass.	229
Prairie sage.	267
B. *DeadPlant.*	
Little barley	156
Downy brome grass	216
Wheat straw mulch	445
Russian thistle	585

percolated into the soil body is retained according to its ca pacity and the rest is lost to subterranean water channels.

Soil moisture relationship. The amount of water greatly fluctuates in the soil and determines the volume of soil gases which are generally inversely proportionate to it, as both occupy spaces in between soil particles.

From the point of view of plant growth and soil moisture-content the texture of the soil is of great importance. The size of the soil particles and their shape, and the size of pores, in the soil body determine the amount of moisture held up in the soil.

Sand which has fairly large particles is the least complex soil and can hold water loosely in the bigger pore spaces. The water generally runs down through it rapidly, percolating to deeper layers. Sandy soils, both physically, chemically and biologically are inert and being able to hold very small amounts of moisture, are very well drained. Clay soils, on the other hand, have a large proportion of colloidal particles, which swell up when wetted and clog pore spaces, thereby making the soils impermeable. The particles of clay are flat, plate like and have large surfaces considered in proportion to their small volumes. These are lamellate and retain high amounts of moisture. Clayey soils are often poorly drained and contain low amounts of gases. Silts which have a large proportion of clay have properties, similar to clayey soils. Loams which are mixtures of sand and clay have the properties of both and are, therefore, most suitable for plant growth. On the whole, loamy soils are very fertile, having fairly large amounts of water and minerals.

The humus particles are hydrophilic colloids. They too are capable of absorbing and retaining large quantities of water. Further, they along with the micro-population and fungi have the property of cementing together clay particles. The aggregation of clay particles improves the soil by opening it, i.e., by providing larger pore spaces for gases. Thus, a clayey soil mixed with humus is very fertile.

The relative proportion of these particles is not only important for the soil-plant-water relationships but on account of the differential swelling properties of the various constituents, they determine the pattern of drainage in an area. If in a soil body a layer of clay lies underneath a layer of sand, water collects in the subsoil and may be available to the vegetation. The thickness of the various layers and their depth from the surface determine whether a particular plant species or community will flourish there or not.

Soil moisture and Indian trees. A great majority of Indian forest tree species grow best on well drained soils, but they can tolerate and even form communities on slightly badly drained soils, which are heavy and clayey. In sal forest of the Dun valley, *Shorea robusta* generally grows best on the Siwalik clay, though it occurs also on sandy soils derived from sand rock. *Terminalia tomentosa* and *Eugenia jambolana* occupy deeper and moister spots in the area. The *chir* pine (*Pinus roxburghi*) occurs on sand, rock or conglomerate, that is well drained. Soils with different textural and drainage pattern bear distinctive flora.

Classification of soil water.—Water occurs in the soil as:—

(a) Gravitational water—occupying large pores of the soil.

(b) Capillary water—occurring as films round the soil particles and in between small pores.

(c) Hygroscopic water—held up in the soil in the form of very thin films by surface forces on the particles.

(d) Water vapour—existing in the soil atmosphere.

(a) **Gravitational water and ground water.** In a heavy rain the larger pores of the soil are filled with water, which percolates to the lower layers by the force of gravity. In this displacement process air is expelled from the soil body and if this water remains for a long time it creates stagnation or waterlogged conditions. When it reaches the hard bed-rock or a hard pan layer, it becomes *ground water;* and its movement is then controlled by the under-ground slope or the strata of the rock. Usually the gravitational water is drained off very soon after a heavy shower from the upper layers of the soil; but if humus and vegetation are present this is held longer by the spongy structure. If this water remains in the soil for a very long period, it becomes injurious by inducing anaerobic conditions for the proper growth of root systems of a majority of plants.

The main factors which control the formation of ground water are slope of the surface,. the amount and rate of rainfall, the temperature of the air, the structure and composition of the rock and the nature of the vegetation. In its initial stages the percolation of water depends upon the porosity of the soil.

Soil erosion and other factors that disturb the porosity of the soil adversely affect the formation of ground waters.

In addition to porosity the quantity of ground water is determined by specific rententive power of the rock. Clayey and shaley rocks can soak larger amounts of water, but they may still be unproductive as this water is tenaciously held up by colloidal matter in the rocks and is not available for plant growth.

Depending upon their porosity and premeability rocks may be classified in the following two categories :—

(i) Water yielding rocks, which are pervious and retain water throughout their mass.

(ii) Non-water yielding rocks, which are nearly impervious in mass, but hold water in joints, cleavages, bedding planes, cracks, fissures, etc.

A rock or a formation that yields water is called aquifer. Sand has a porosity of 40%, newly deposited alluvial sand has 10% porosity as against 10-15% of sandstones. Clays may absorb water equal to nearly 52% of their volume, but this water is firmly retained in it. The strata of the clay when interbedded with the strata of sandstone create impervious surfaces, and in such layers water percolates laterally to long distances and comes on the surface in fissures or cracks, becoming available for plant growth only in springs or rivulets at lower levels.

From the point of view of water supply, Auden (1940) classifies Indian rocks into the following broad categories :—

(1) "Unconsolidated gravels, sands, muds and cl ays in which there are abundant pore spaces or voids are usually filled by water when the rocks are below the level of the water table." These rocks occur in deltas of most of the rivers. in the Indo-Gangetic alluvium, in stream beds and in valleys of

rivers that may have lakes and in beeches and dunes near sea coast. In such regions, water is available for plant growth in fairly large quantities.

(2) "Consolidated conglomerates, sandstones, shales, chalk and soft laterite in which pore spaces generally occur, although they have been reduced in volume by pressure and compaction, and also by deposition of an interstitial cement of silica, lime and iron oxide. Water occurs in these reduced pore spaces, and also along joints and bedding planes." For plant growth water is not available in sufficient quantities in these rock types, as after rainfall it quickly percolates into the soil. These rocks are therefore dry, even if they occur in regions of high rainfall; and usually bear a dry type of vegetation.

(3) "Quartzite, limestone, slate, schist, gneiss, granite, basalt—Some of the rocks in this group have been still more compacted by heat, pressure and deposition of silica, lime, etc., with the result that the pore spaces are either very small, or completely absent. Other rocks in this group, having solidified from a molten state, never possessed pore spaces. Water is found in such rocks along joints, bedding planes, fissures and also along fault and crush zones. Water is also frequently found in soft porous patches on the ground surface where granite and gneiss have decayed and disintegrated by the processes of weathering." For the purposes of plant growth these rocks provide rather poor surfaces, unless they are finely weathered; in which state they retain fairly large amounts of moisture.

In areas where these rocks are horizontally stratified the surfaces available for plant growth are very poor in moisture, as a result of which they bear a shallow rooted scrubby vegetation, or a poor type of grassland. In valleys excavated in such formations the two sides of the valley do not show much differences in vegetation; and whatever little differences are there, they have been produced by variations in temperature and evaporation conditions, coinciding with aspect or slope. The examples of this type occur in the Deccan Peninsula, where valleys bear luxuriant tropical vegetation, more or less uniform throughout, on account of the favourable factor of the rainfall and temperature.

On deforestation the amount of water in the soil becomes very scanty and afforestation of such areas becomes extremely difficult on account of the non-moisture-retaining properties of parent rocks.

In the Himalayas and such other areas, where sedimentary rocks belonging to the groups 2 and 3 occur tilted to one side or the other, interesting water relationships on the two slopes in valleys are seen. On account of the percolation of the water through the bedding planes, joints, etc. the scarp slopes remain dry, even after a heavy rainfall. On the other hand, dip slopes at lower levels receive abundant seepage water. In Fig. 23 the effect of aspect and geological slope on the two sides of Himalayan valleys is seen at many places. The scarp slope in addition to being physiologically dry receives higher amounts of sun's heat as well. Consequently, more vegetation will result on the dip slopes as mesophytic type of compared to xerophytic plant cover on the escarpment. If the

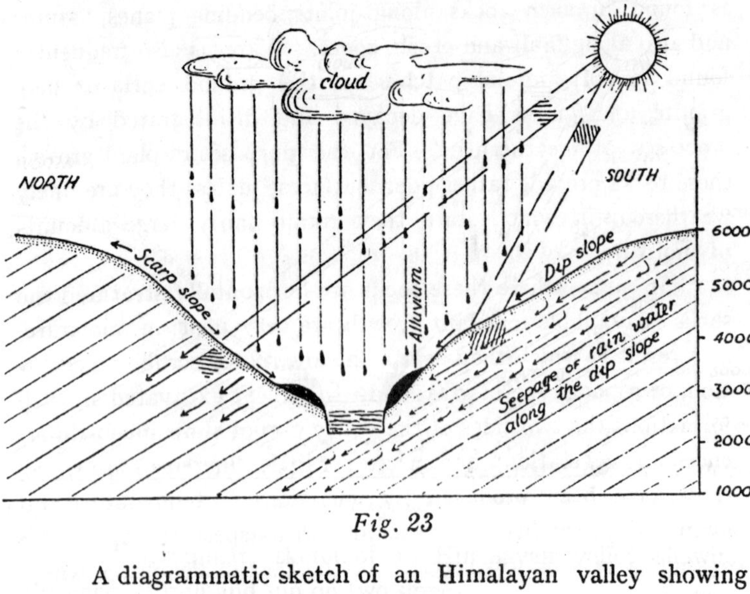

Fig. 23

A diagrammatic sketch of an Himalayan valley showing the effect of rain and sun on two slopes.

rocks are rich in bases the percolating waters will carry dissolved in them mineral salts which will be available only on the dip slopes. Due to the presence of greater amount of moisture on the dip slopes, the temperature here will be more uniform than on the scarp. This will result in the development of an exacting type of vegetation on the dip slope and a non-exacting type on the scarp.

Examples of such variations in the development of vegetation due to differences in water absorbing and water retaining properties of various rocks and rock formations mentioned above are seen abundantly in the Himalayas (see Puri, 1950). In Fig. 24 the types of vegetation growing on scarp and dip slopes are shown from the Parbatti Valley in the Kulu Himalayas.

The inclination of rocky strata produces more favourable effects for the development of vegetation at various places on the dip slopes in regions with low rainfall, usually below 80 inches. But this effect has been shown by Stamp to be occurring even in those forests of Burma, which have rainfall higher than 80 inches.

If the strata of the rock dip below the surface of the ground and no fissures, or cracks occur above ground, the water percolates into the soil and is lost so far as the plant growth is concerned. This subterranean water rises in springs and rivulets. Vegetation around such spring heads and rivulets is usually greener and shows distinctly the effect of flushes.

As subterranean water is one of the important factors for plant growth in this region it will be useful to add something on its occurrence in springs. Springs are classified into the following types: —

(1) Gravity springs—are those in which water comes on the surface at lower levels. These may be either depression springs, contact springs or fracture and tubular springs.

In depression springs, water appears on the surface from permeable lower layers, because the surface descends below the water table. In contact springs water comes on the surface "from permeable material, over the outcrop of less

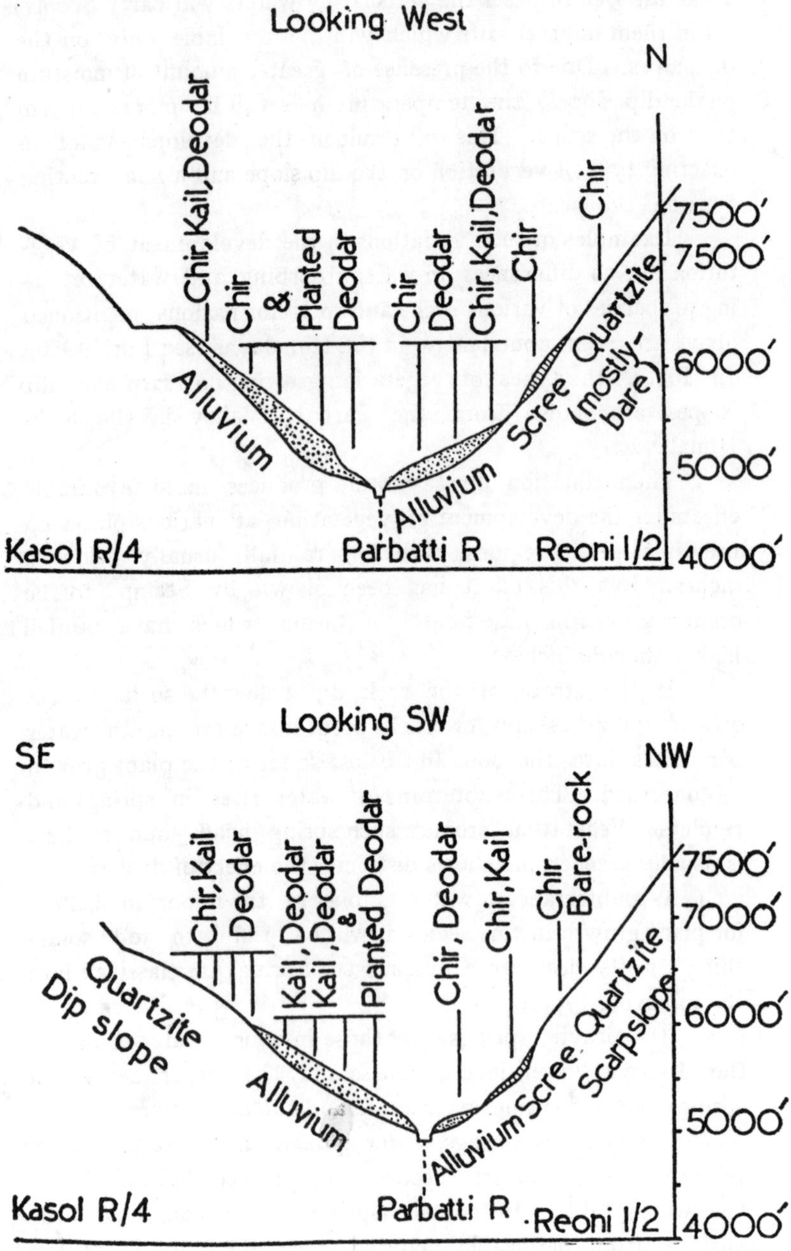

Fig. *24*

Relation between tree vegetation and geology in the Parbatti Valley, Kulu Himalayas.

(a) Section a little below Village Chhalal.

(b) Sction above half a mile from Village Chhalal.

or impermeabe material, which retards or prevents the downward percolation of the ground water and thus deflects it to the surface" In fracture and tubular springs water flows through openings in the rocks.

(2) Artesian springs—in which the water of a permeable bed, or a fissure, rises to the surface by hydrostatic pressure, after confinement between two impermeable beds or surfaces.

(3) Springs that may be associated with volcanic and other activities in the surface of the earth.

Ghose has recently classified the following four major tracts in which great majority of the Indian springs occur:

(1) Himalayan belt; (2) Sind—Baluchistan belt; (3) Bihar belt and (4) West coast belt including Bombay Presidency.

The depth of water table influences very greatly the type of vegetation that develops in a region. In monsoonic climate it will be of great importance to know exactly the depth at which water table occurs, so that the correct use of land be assessed in conservation and land management. This will also tell whether a particular area could be effectively used for afforestation purposes and the type of species that would be most suitable. For example, in regions with very low water table, species with long tap root systems will only be successful, but in regions with high water table species of spreading root type will ordinarily flourish better.

Coulson (1940) states that "the depth of the water table below the surface of the ground is very variable for it depends on such widely different factors as topography, geology and climate. It rises and falls with the general undulation of the country, but with smaller variation with height; it is deepest below high peaks and in regions drained by deep valleys. In some places it may be hundreds, or even thousands of feet below the surface. The inclination of the water table is modified not only by the slope of the ground but also by differences in the soil texture, and in the character and position of the underlying rocks and vegetation; the water table thus lies nearer the surface in fine textured soil than in coarse; and nearer under tilled fields than under forests, where it meets the surface of the slope, ground water would drain out and give rise to seeps and springs and where it rises above the floor of a confined hollow it would form ponds." During rains therefore, water table rises and falls in periods of drought.

The rate of flow of ground water may be between 4 ft. to 10 ft. a day, but in the Indo-Gangetic alluvium, which is composed of alternating beds of clays, sands, loamy layers, with the slope for the movement of the ground water nearly 1 to 250 the speed of the percolation of the underground water does not exceed more than a mile per year. These deposits are very deep and it is held that beneath these the strata of the Siwaliks, occur dipping at various angles.

Auden (1940) has divided India into three main ground water provinces, which are given in Table 28.

Thus, conditions for the growth of vegetation in different parts of India are very different and ground water provinces combined with rainfall intensity zones may give better relationship in deciding the various vegetation regions of India.

It may be considered that, in regions with inclined strata of the rock, especially in the Himalayas, some types of vegetation are influenced mainly by the water percolating through the strata of the rock; and their relationship with rainfall is very remote. Some of these may give the appearance of climax types. Examples have been given by Puri (1950) from the Kulu Himalayas. It is, however, extremely difficult to determine their ecological status. Whatever they may be they are certainly not climax types, but only specialised seral stages, preserved in their present state by the special features of the habitat.

(b) **Capillary water**—When gravitational water has percolated to the lower layers of the soil after a shower the remaining moisture is held as thin films around the soil particles and in the angles between the particles and in smaller pores. At this stage the soil is at its field capacity and sufficient water being held up loosely, is available for plant growth.

Ground water provinces of India.

(After Auden, 1940).

Province.	Sub-Province.	Principal manner of occurrence of ground water.	Recovery of ground water.
Northern montane zone.	Mountains.	Run-off high, except under snow. Percolation in joints, bedding planes, scree gullies, surface-creep material.	Springs.
	Valley-fill deposits.	Pore-spaces of gravels and sands.	Tube-wells, percolation wells, etc.
Indo-Gangetic alluvium.	Ganges-basin.	Abundant quick-sands. Water table high. Ground water mostly fresh. Some areas of sodium carbonate and sulphate efflorescence.	Tube-wells, percolation wells.
	Lahore basin.	Water-logging towards south-west and probably quicksands. Water table low towards north-east, soils generally more alkaline than those of the Ganges basin.	Tube-wells, percolation wells.
	Lower Indus basin.	Water table low and in some parts over 300 feet below ground level; saline water generally.	Deep percolation wells. In regions where water table raised by infiltration from rivers, by tube-wells infiltration galleries.
Peninsula.	Archaean.	Water in joints, fissures, crush zones, weathered gneiss.	Percolation wells with infiltration galleries;. some drilled wells also.
	Older sedimentaries and Deccan traps.	Most of rocks with low porosities. Water in joint and bedding planes.	Percolation wells and springs.
	Younger sedimentaries.	Rocks with higher porosities. Water held partly in the body of rocks and partly in bedding and joint planes.	Percolation wells. Tube wells in rocks with sufficient porosities and transmitting properties which are not saline; and in springs.
	Coastal and river alluvium.	As in Indo-Gangetic alluvium.	Tube-well, etc.

The moisture content of finer soils at field capacity is higher but much of it is unavailable to the plant being locked up as hygroscopic water around the colloidal matrix.

The movement of capillary water takes place in the soil by capillary action, chiefly in the vertical direction, and lateral movements of capillarity are rather small. Kramer (1949) states that if "these forces are expressed in terms of tension" water moves along a gradient from a region where the forces holding water are low to one where they are higher, in other terms, from a region of low pF to one of high pF, from a region of high vapour pressure to one of lower vapour pressure, or from one of low diffusion-pressure deficit to one of high diffusion-pressure deficit. Regardless of the terminology, water always moves along a gradient of decreasing free energy".

As the amount of free energy for the movement of moisture is considered to be high in moist soil and low in dry soil water movement is high in sandy soils and low in clayey soils. But if the soils are dry then the movement is slow in sandy soils and rapid in clays. Recent work has shown that capillary water moves slowly from moister to dry parts of the soil, if the water table is lower than 3 or 4 ft. Unless the soil is saturated at one end the movement of moisture is very slow.

Some experiments conducted in the U.S.A. have shown that in dry soil water moved only 8 inches in 139 days. In another set of experiments it was shown that "a period of 180 days was required for all the soil in a 6 inch auto-irrigator part to be wetted from the air-dry condition at a condition of equilibrium with a tension of only 150 cm. of water." This may show that in dry period when transpiration is high the available water surrounding the roots in the soil particles is lost more quickly than it is replaced by capillary movements in the soils, as a result of which a cylinder of dry soil may surround the roots and cause them to wither.

Gupta has recently stated that the survival of *Casuarina* plants in some of the plantations in the Madras State is related to the movement of capillary water in sandy soils. A correct determination of the maximum available water for free seedling-growth which is the difference between field capacity and wilting coefficient will yield important data in the afforestation of desert areas. In a bulletin entitled the "Rajputana Desert" (1952) some data have been produced concerning this point.

(c) **Hygroscopic water.**—The water which remains in the soil after capillary water has been removed by air-drying is called hygroscopic water. It is mostly held as very thin films on soil grains and colloids. When air-dry soil is kept in contact with moist air it absorbs moisture and it becomes difficult to distinguish sharply between capillary and hygroscopic water. Hygroscopic water is obviously unavailable to plants and is also known as bound water.

(d) **Water vapour**—is formed in the soil atmosphere and is in equilibrium with liquid water, temperature and pressure. It moves in the soil along vapour pressure gradients and as it may condense to form liquid water it is of importance in soil-water relationships.

As the movement of the water vapour is along vapour pressure gradients it is affected by changes in temperature and vapour pressure of the various layers of the soil and of the soil surface and the air directly in contact with it.

The importance of water vapour for plant growth is immense in arid regions where there is no water table. Extensive studies of Lebdeff (1928) in Russia have shown that during winter fairly large quantities of water vapour move from lower warmer layers of the soil and condense in the upper cooler layers. In summer when the upper layer of the soil is cool during the night, the water vapour moves to the upper layers, but during the day when the upper layer becomes comparatively warmer there may be some downward movement of water vapour as well. This diurnal variation in soil temperature may be responsible partly for supplying moisture for the growth of desert plants.

Wilting and availability of water.—It must be however, clearly understood that the distinction between gravitational, capillary and hygroscopic water is only qualitative and that intermediate stages make them continuous. A plant will continue absorbing water as long as the forces of absorption exceed those with which water is held by the soil. During the rainy season in India the soil may be saturated or remain at field capacity, for most of the time. Soon after the rainy period shallow rooted plants suffer from a water deficit. In the day time when the rate of transpiration is high and the rate of absorption cannot keep pace with it the plants temporarily wilt just to recover in the night on reduction of water loss. However, a time comes when the soil has become so dry that the physiological equipment of the plant has been fully taxed (such as the rise of osmotic concentration of cell sap) for any further absorption. The plant permanently wilts at this stage. Soils differ in their water content at this time but most of the plants with similar root system wilt permanently on the same soil. That is why a large number of seedlings and herbs of all species growing together wither away in the post-monsoon season.

The range between the moisture content of the permanent wilting point and that at the field capacity is wider in finer and clayey soils than in sandy soils. Since this is the range of available water finer textured soils can support plants for longer periods.

Water in relation to structure and growth of plants.—The production of root hairs in most land plants takes place if the soil is well aerated, and the amount of moisture is between the permanent wilting point and the field capacity. In dry soils or those which have high contents of dissolved salts, the production of root hairs, and consequently the absorption of water from the soil is inhibited,

The development of root systems depends upon the amount of moisture present in the soil although in most cases it is governed by hereditary factors. If the water is low in surface layers, plants develop long tap root system, which are variously branched. But in localities with high water content near the surface, root system is short and branched. Most of the trees in Indian forests probably belong to the first category of plants. Some Indian trees which produce shallow roots occur either in very moist localities, or in areas with water table close to the surface.

A large number of trees and shrubs belonging to the Abietineae, Fagaceae, Salicaceae, Betulaceae, etc., have two types of roots—long and short. The original root elongates and produces a number of short roots without root caps. It has a fewer root hairs, which are short lived, usually for one season. The long roots have suberised root tips through which some absorption of moisture takes place. The short roots develop mycorrhiza, and fungi of the Basidiomycetes-type develop in them. These mycorrhizal roots increase the absorbing surfaces of roots and are helpful in the absorption of water and solutes dissolved in water.

On account of the fact that below field capacity, the movement of water due to capillary forces in soils is very small, roots continually extend to lower regions of the soil. The survival of tree seedlings is therefore, dependent on the rapidity and ease with which they produce a well branched, long, root system.

In perennial plants, root system penetrates to deep layers of the soil and roots have been found below 30 feet in some species. Some of the species of *Prosopis*, *Salvadora*, *Zizyphus* etc. in north-west India have roots reaching to very deep layers of the soil. In addition to vertical growth, roots develop horizontally and occupy areas larger than those occupied by their crowns.

The growth and development of grass roots and those of crop plants have been studied in greater detail than the forest trees.

Working in the United States, Coile (1937) found that "per square foot of trench face in the top 5 inches of an oak forest, there were as many as 250 roots less than 0.1 inch in diameter, or more than 90% of all roots of that size class occurred to a depth of 36 inches. In a 35 old loblolly pine stand, there were about 160 small roots per square foot of trench face in the top 5 inches, again more than 90% of all the roots of this size." Very little is known of the root development in relation to moisture content of the soil and tree regeneration in Indian forest species.

Root systems of woody plants are generally perennial, but a large number of the smaller roots are supposed to die every year, depending upon the time of the year, when the soil gets filled with water in heavy rains. Here, again, there are no data from Indian conditions.

It has been shown in Europe and America that during the periods of worst drought, the seedlings that suffer the most are those which have shallow root systems. The development of root system in addition to heredity depend upon the environment and root competition is one of the serious factors that may limit the development of roots because roots compete for water and soil minerals. The competition between roots is, therefore, intense in the upper layers of the soil and in a forest stand, when the number of roots present have reached a saturation point, the invasion and establishment of other seedlings is not possible. Some species on account of taking higher amounts of mineral and moisture from the soil in their growth, deplete the upper layers of the soil and thus make the habitat unsuitable for the establishment of their own seedlings. This may also be due to the production by roots of toxic substances. Toxic effect is also produced by the decomposition products of forest litter. Root secretion may also be one of the weapons through which root competition takes place in dense patches of vegetation.

The root competition among plants becomes more intense if limited supplies of soil moisture are present. Generally, in dry soils plants have longer root systems but in very dry soils the elongation may cease and root tips may become suberised. In dry soils, although, roots are not able to absorb sufficient quantities of moisture, nitrogen and potassium may still be absorbed to a moderate extent.

When the supply of moisture in the soils is small, the ratio of root to shoot development is high. It is not always true that roots penetrate to the moister region of the soil as the downward growth of roots from moist to dry soil has also been seen. The inhibition of roots in regions of dry soil is due to dessication.

Excessive water and water-logging.—The amount of water in the soil is related to the amounts of soil air and oxygen. Generally, soil air is inversely proportional to soil moisture. In soils with high contents of water, therefore, dissolved oxygen is present in very small amounts. As a result of this, normal oxidation processes related with metabolism of plants stop in water-logged soils and reduction processes take place, instead.

Various methods are designed to measure these processes in water-logged soils and the most usual of these is the measurement of the dissolved oxygen in water. This, however, failed to indicate exactly the point at which oxidation stops and reduction starts, or, *vice versa*. The most accurate and easy method suggested by Pearsall (1938) and Pearsall and Mortimer (1939) is to measure the oxidation-reduction potential of soils and muds and relate it with the associated biological processes. Pearsall and Mortimer state that "in all cases so far examined in soils, water and muds, representing a wide variety of natural conditions products of oxidation (ferric iron, nitrate, sulphate) are found above the potential zone E5 + 350 mV.; below this potential their reduced counterparts (ferrous iron, ammonia, sulphide) are present—of great ecological interest is the demonstration that the change over from reducing to oxidizing conditions (at E5 + 350 mV.) in natural water systems takes place at relatively low oxygen concentration (8% saturation or less). From this, these authors inferred that fairly low amounts of oxygen in natural soils can maintain oxidation processes.

Pearsall has related these changes of redox potential with the availability of plant foods and with the distribution of various ground flora communities and seedling growth in English woodlands in Table 29. It will be apparent that below Eh 355 mV. generally, nitrates are not present in the soils. These soils have a ground flora distinct from the oxidizing soils, in which nitrate is present.

TABLE 29

Variations in soils redox potentials, bases and nitrates in soils of the given plant communities.

(After Pearsall, 1938).

Vegetation.	pH	Eh	E5	Thyiocynate Alone	test. H_2O_2	Nitrate
Carex inflata.	5.38	130	152	1	3	...
C. inflata, Menyanthes.	5.47	155	182	1	2	...
Myrica, Menyanthes, Sphagnum,	5.50	158	187	...	2	...
Salix cinerea.	5.68	10	48	1	3	...
S. cinerea.	5.49	135	163	1	3	...
S. cinerea.	5.18	314	325	2	3	...
Erica tetralix, Myrica, Sphagnum.	4.30	275	234	2	3	...
Erica tetralix, Myrica, Sphagnum, Calluna.	4.13	260	211	2	3	...
Eriophorum vaginatum, Sphagnum,	3.85	355	288	1	2	...
E. vaginatum, Erica tetralix.	3.81	345	276	1	2	...
Spiraea ulmaria, Phalaris arundinacea.	6.28	325	399	2	2	2
Juncus sylvaticus	5.08	655	360	2	2	1
Molinia coerulea, Lythrun salicaria.	6.01	300	358	3	3	1
Teucrium scorodonia	4.35	505	467	3	3	2
Pteridium aquilinum.	4.04	440	384	3	3	1
Mercurialis perennis.	5.15	340	349	1	1	3
Pteridium aquilinum.	4.54	445	418	2	2	1
Brachypodium sylvaticum	5.96	425	481	...	...	2
Urtica dioica.	6.58	395	487	...	...	2
Pteridium, Holcus mollis.	4.26	630	587	3	3	1
Mercurialis perennis.	5.22	450	463	1	1	1
M. perennis.	7.68	350	506	...	...	1
Brachypodium sylvaticum.	7.21	370	506	...	...	3

Pearsall has further presented a classification of oxidizing and reducing soils on the availability of plant foods and the growth of characteristic plants, which is given below:—

A. **Oxidizing soils.**

1. Base deficient soils lacking nitrates and with a pH below 3.8 in England, characterised by communities of *Vaccinium myrtillus*, *Dicranum majus*, *Deschampsia flexuosa*, brichwoods and birch seedlings.
2. Base-deficient woodland soils (usually oakwoods) between pH 3.8 to 5.0 (approx.) with nitrates (and earthworms) normally present characterised by*Holcus mollis* and *Milium effusum* and general oakwood (defined) communities with oak seedlings.
3. Woodland soils above pH 5.0 (approx.) with nitrate present, calcareous to slightly base deficient, characterised by *Mercurialis perennis*, *Brachypodium sylvaticum* communities and ash seedlings.
4. Soils oxidizing in summer tending to be reducing in winter, pH above 3.8, nitrates present in summer, characterised by *Alnus glutinosa* woods and probably by *Phalaris arundinacea* and *Spiraea ulmaria* communities.

B. **Reducing soils.**

pH tending to be above 5, nitrates absent, characterised by *Salix cinerea* woods.

Misra (1938) further showed that waterlogged muds when exposed to air become acidic and rapidly oxidising. Rapid drifts in the redox potential on waterlogging and exposure is considered by him as a measure of fertility. Peaty and sterile soils do not show much drift.

The physiological effect of flooding on plants. In a series of experiments on tomato, tobacco, sunflower and yellow poplars, Kramer (1951) has shown that flooding with water of potted plants resulted in almost all cases injury to the plants. It was usually assumed that the wilting of leaves as a result of flooding was due to failure of absorption of water, but recent work has shown that the dying roots of plants produce toxic substances.

Evidence has been brought forward by several workers to show that with the death of root system of plants after flooding, there develop adventitious roots from the stem portion. These begin absorbing water from the medium and make some plants revive. The development of roots is considered to be due to the disturbance of translocation of carbohydrates and also hormones.

Another cause of the death of roots in flooding is the decrease in amount of dissolved oxygen which disturbs the normal oxidation processes. In anaerobic conditions the compounds formed are nitrites, ammonia, sulphides and ferrous iron due to reduction. These are toxic to roots of most plants and there results a state of deficiency of plant foods and minerals, which probably kills the plants.

The root systems that are developed in anaerobic media are physiologically and anatomically distinct from those that are developed in well aerated soils or water cultures. It has been reported that roots of barley, corn, oats, tomato, etc., developed in unaerated water cultures have a large number of large air spaces in the cortex and some of them were less suberized than those roots that were developed in aerated cultures. The reasons for the death of root systems of plants when flooded are probably really due to their being physiologically different from roots developed in aerated media. On account of the resultant anaerobic conditions due to flooding these are more probably unable to absorb food material and eventually die.

The above review of the present knowledge of the problem indicates that we do not as yet know fully the reasons for the death of plants after flooding.

Ecological classification of plants based on their water relationships. Depending upon the amount of water present in the soil, Warming (1909) gives the following classification of plants, occurring on soils of various water content.

1. *Hydrophytes* are the plants which occur in water.

2. *Halophytes* are the plants occurring in marshy places.

In these two types the soil is extremely wet, and water is available to plant growth in abundance. The tissues of these plants contain generally large lacunae, or intercellular spaces. On account of the presence of air in these cavities, the plants are usually floating on the surface of the water. Aerenchyma may be equally developed in roots, shoots or fruits of most of the aquatic plants. Cuticle or periderm is absent and stomata are present only on surfaces exposed to air. Cutin and suberin are not formed in the parts of the plants, which are submerged. Absorption of water takes place through all the submerged parts although root absorption is considered to be generally essential for the best growth of water plants. Roots are, on the whole, shorter, and much less branched, as compared to terrestrial plants. Root hairs are not well developed and root caps are sheath-like and elongated.

According to the way in which hydrophytes are developed in water, the following four chief categories may be recognised:–

(a) Free floating hydrophytes are exemplified by *Lemna, Spirodella, Wolffia, Salvia, Azolla, Pistia*, etc.

(b) Rooted submerged hydrophytes are *Vallisneria, Zostera, Hydrilla, Potamogeton, Ceratophyllum, Ottelia*, etc.

(c) Rooted hydrophytes with floating leaves are *Nymphaea, Nelumbium, Limnanthemum, Aponogeton*, etc.

(d) Rooted and immersed occurring in shallow water are *Typha, Sparganium, Scirpus, Eleocharis*, several species of *Polygonum*, etc.

3. *Mesophytes* are those plants, which require a humid atmosphere for their growth. Normally, they are incapable of growth on wet soils, or in water and require well aerated soils. The plants are usually large-leaved and possess certain characteristics of hydrophytes. Broad-leaved forest trees, which occur in wet depressions or along lakes and rivers on alluvia such as *Juglans, Aesculus*, Poplars, Cherry, etc. are usually cited as examples of mesophytes. Incidentally, these plants have a high amount of foliar Ca and nitrogen and C/N in these is quite low.

4. *Oxylophytes* occur on physiologically dry and acid soils.

5. *Psychrophytes* are the plants that occur on cold soils.

6. *Halophytes* occur on saline soils.

The three above types of plants occur on physiologically dry soil, and their growth is dependent on thier capacity to absorb water, the effect of climate being secondary.

7. *Lithophytes* occur on rocks and stones.

8. *Psammophytes* occur on sands and gravels.

These two types occur in very dry soils, and in localities where climate is exceedingly dry. Types 4-8 are Xerophilous. Although, it is difficult to give an exact definition of a Xerophyte, the one given by Daubenmire may be taken as a standard. He states that "Xerophytes are plants which grow on substrata that usually become depleted of growth water to a depth of at least 2 decimeter during a normal season. In arid regions all plants not confined to the margins of streams or lakes are considered xerophytes, whereas in regions of high rainfall the class would be represented only by shallow rooted plants of sandy soils, by plants of dry ridge tops, by Algae, mosses, and lichens which grow on tree bark or rock surfaces, etc."

Xerophytes usually possess certain structures in their leaves by which the loss of water from plant surfaces is greatly reduced. Morphologically, the leaves may be greatly reduced, needle-shaped with reduced lamina or may be variously covered with woolly hairs, etc. Structurally, the stomata may be sunken, and the cavity may be bounded by thick walled cells which have cutinized walls. In ***Saccharum munja*** the leaves are rolled with stomatal surfaces inside. In still other cases, cuticle is thick and stomatal chamber may be blocked by resin or wax.

The subterranean organs may be transformed into a long taproot which would facilitate the maximum absorption of water from the lower layers of the soil. In some plants roots are fleshy, tuberous or bulbous.

The plants of deserts and cold regions are capable of surviving a prolonged period of physical or physiological drought as their leaves are shed and their buds are covered variously.

In some plants leaves get rolled up during the periods of drought; in other cases, fleshy leaves or stems which retain higher amount of water are developed. Succulent phylloclades of *Opuntia* do not loose water readily if placed to dry on removal from the plant. The pentosans and other hydrophilic colloids retain and conserve water. Sclerophyllous tissue may be developed in some leaves and in others both upper and lower epidermis are thick and mesophyll cells are compact.

The leaves of xerophytes have usually high osmotic pressures by which a considerable amount of suction pressure is developed and the absorption of water from the soil is effected. The high concentration of the cell sap is responsible for deccreased transpiration. The role of osmotic pressure in the water economy of Xerophytes is not clearly understood, however, it is now considered by most workres that in drier regions, the lesser the amount of water in the soil, the greater will be the osmotic pressure in plants occurring in such localities.

The osmotic pressure of a few plants are given in Table 30 on p. 159.

The data show that high osmotic values (0.5—0.9) are found in xerophytes and lowest values (below 0.2) are found in succulents. In case of halophytes the values are very high.

The true nature of Xerophytism is not clearly understood. For example, it is difficult to decide whether a xerophyte is really xerophyllous and occurs only in dry habitats and deserts, or it is merely drought resistant. It is observed by many workers that desert plants when grown in soils with higher amounts of water show better growth. Experiments on *Artemisia fasciculata, Zygophyllum fabago* and *Centaurea solstitialis* show that when water holding capacity of the soil was increased from 40 per cent to 60 per cent the yield of the plants also increased to 1½ to 2 times. Maximov (1929) thus concluded that "xerophytes are not xerophyllous, at least so far as the soil moisture is concerned. Xerophytes develop better with abundant water in the soil, and therefore, must be reckoned not as drought resistant plants."

TABLE 30

Osmotic values in the leaves of plants according to Maximov, Dilanian and Silikova.

(After Maximov, 1929 and Navalkar, 1940).

Species	Osmotic value in atmospheres
(a) Xerophytes:	
Artemisia maritima	0.50
Gypsophila acutifolia	0.50
Kochia prostrata	0.60
Centaurea ovina	0.70
Parietaria judaica	0.80
Zygophyllum fabago	0.80
Dianthus fimbriatus (=*D. orientatis*)	0.90
(b) Mesophytes:	
Erodium ciconium	0.30
Papavera strigosum	0.30
Hirschfeldia adpressa	0.30
Senecio vernalis	0·40
(c) Succulents:	
Sedum maximum	0.15
Sedum oppositifolium	0.15
(d) Halophytes:	
Avicennia alba	31.01

The sap of ecologically distinct plants possesses different physico-chemical properties. The osmotic pressures of leaf sap of hygrophytes, mesophytes and xerophytes are given in Table 31.

TABLE 31

Osmotic pressure of cell sap from leaves.

(after Maximov, 1929).

	Osmotic pressures of leaves in atmospheres.	
	Herbs	Woody plants
1. *Hygrophytes:*		
Jamaican montane rain forest	8.80	11.44
2. *Mesophytes:*		
Long Island	10.40	14.40
3. *Xerophytes:*		
(a) Arizona desert	15.15	24.97
(b) Jamaican coastal desert	...	30.05

It is clear that xerophytes have higher osmotic pressures in the cell sap and woody plants, within the same class, have on the whole higher cell sap pressures than the herbaceous vegetation. This is further seen from Table 32.

TABLE 32

Osmotic pressure of cell sap from leaves.

(After Maximov, 1929).

	Atmospheres
Trees and shrubs	28.10
Dwarf shrubs and undershrubs.	21.45
Perennial herbs	16.35
Winter annuals	14.75

Like pH and other soil features, plants have definite soil moisture tolerance. Many plants like teak and sal are very sensitive to higher moisture relations. When waterlogging conditions are about to set in, these two species show signs of degeneration. There are others like *Eugenia jambolana* and *Lagerstroemia parviflora* that occur in dry as well as in very moist soils. No work has so far been done in India, on finding out the soil water relationships of plants, but data for some American species are given in table 33.

TABLE 33

Relative abundance of given species of plants in water holding capacity classes of soils studied in New England, U.S.A.

(After Colvin and Eisenmenger, 1943)

Plant.	Soil horizon	Abundance of plants in water holding capacity classes of soil.										Mean water holding capacity of soil.
		Low			Medium				High			
		30-40 %	40-50 %	50-60 %	60-70 %	70-80 %	80-90 %	90-100 %	100-110 %	110-120 %	120-130 %	%
Lupinus perennis. ...	A	67	33									38.3+ .47
	B	95	5									35.5+ .22
Lespedeza capitata. ...	A	30	48	12	10							45.2+ .91
	B	87	9	4								36.7+ .47
Baptisia tinctoria. ...	A	70	22			8						40.4+ 1.1
	B	57	23				20					47.3+ 1.93
Quercus ilicifolia. ...	A	60	23	6	9			2				42.4+ 1.21
	B	80	14	5	1							37.7+ .60
Pinus rigida. ...	A	55	29	11	4			1				41.9+ .99
	B	74	13	1	1			1				39.4+ .91
Andropogon scoparius. ...	A	18	48	22	11	1						47.9+ .92
	B	65	26	9								39.4+ .65
Gaylussacia baccata. ...	A		29	15	6	46	4					63.1+ 1.38
	B	33	24	4	18	21						52.0+ 1.58
Prunus virginiana. ...	A	36	12	17	8	13	5	7	2			55.3+ 2.06
	B	16	29	17	13	11	7	7				57.3+ 1.78
Juniperus virginiana. ...	A	2	25	15	9	35	10	4				64.6+ 1.56
	B	8	57	29	6							48.31+ .71
Betula lenta. ...	A	5	9	4	23	21	16		17	5		75.9+ 2.12
	B	9	16	23	11	37	4					61.3+ 1.44
Betula alba var. *papyriferia.*	A	4	2	2	8	8	6	13	34		23	96.5+ 2.40
	B	5	9	11	18	20	37					70.0+ 1.53
Acer sccharum. ...	A	2	1	2	9	10	13	17	32		14	93.9+ 2.05
	B	1	10	16	19	22	25	7				70.4+ 1.49
Quercus rubrum. ...	A	3	7	6	11	10	13	19	29		2	84.2+ 2.15
	B	4	6	16	21	18	18	17				71.5+ 1.66
Acer rubrum. ...	A	3	6	5	16	8	9	14	14	12	13	88.6+ 2.57
	B	1	10	25	8	20	14	22				71.6+ 1.75
Viola pedata. ...	A		40	53	7							51.7+ .60
	B	92	8									35.0+ .27
Hamanelis virginiana. ...	A		11	5	24	8	15	16			21	83.5+ 2.42
	B	4		12	25	33	26					71.1+ 1.22
Alnus incana. ...	A				18	47	15	12			8	81.1+ 1.56
	B		8	13	29	42	8					67.9+ 1.05
Fraxinus americana. ...	A			10	8	6	10	21	32		13	88.3+ 1.63
	B		1	16	21	35	20	7				72.8+ 1.17
Fagus grandifolia. ...	A			1	4	10	9	9	39		28	120.8+ 1.82
	A		1	10	15	60	14					78.0+ 1.11
Potentilla fruticosa. ...	A					4	16	1	16	19	44	111.2+ 1.61
	B		5		19	5	25	46				83.3+ 1.43
Picea canadensis. ...	A				14	6	16	19	37	3	5	93.8+ 1.61
	B			1	2	64	20	13				79.3+ .78
Cornus florida. ...	A				23	62	15					74.2+ .61
	B			38	43	10	9					64.0+ .91
Ulmus americana. ...	A			1	1	7	20	13	24		4	103.5+ 1.82
	B		23	8	17	18	34					68.2+ 1.56
Taraxacum officinale. ...	A					10	17		50		23	103.2+ 1.57
	B		11		43	22		24				72.2+ 1.51

SOIL SOLUTION

On account of the great dissolving power, water is seldom found in a pure state. The solid matter of the soil solution serves as a store house for nutrients and foods of plants. Complex mixtures of inorganic substances, as carbonates, sulphates, nitrates, chlorides and organic salts of calcium, magnesium, sodium, potassium, etc. may be found in water. In addition, free ammonia, albuminoid ammonia, oxygen, carbon dioxide, nitrogen and nitrates and other substances of organic origin may also be present.

River and stream water.—The chemical composition of water issuing in springs and in rivers is determined by the nature of the rocks through which water has percolated and the climatic condition of the region, chiefly temperature and rainfall. The composition of mineral matter of James river and Pecos river in New Mexico, which are in regions of high rainfall and low rainfall, respectively, is compared in Table 34.

TABLE 34

Composition of the mineral matter of rivers from regions of high and low rainfall.
(After Browne, 1938)

Ingredient.		Water of James river (High rainfall). %	Water of Pecos river (Low rainfall). %
Carbonic acid.	...	42.52	1.54
Sulphuric acid.	...	5.26	43.73
Nitric acid.	...	3.32	—
Phosphoric acid.	...	Trace	—
Chlorine.	...	1.51	22.56
Calcium.	...	18.49	13.43
Magnesium.	...	5.44	3.62
Sodium.	...	3.52	14.02
Potassium.	...	3.58	0.77
Silica.	...	14.74	—
Alumina.	...	0.58	—
Iron oxide.	...	0.96	0.33
Manganese oxide.	...	0 08	—
	Total	100.00	100.00
Total solids:--			
Parts per million.		69	2,834
Percentage.		0.0069	

It follows that the amount of mineral matter present in a river depends on (1) composition of the rock through which water percolates and (2) the climatic conditions of the area.

The high rainfall river water has larger percentage of carbonic acid, while higher amounts of sulphuric acid is reported from the river from low rainfall region. The latter has higher amounts of chlorine and sodium salts, while the former has higher amounts of silica and calcium salts.

Some analyses of Brahmaputra river water near Tezpur in Assam are presented in Table 35.

TABLE 35

Analysis of Brahmaputra water at Tezpur.

(After Coulson, 1942).

Number of Analysis.	1	2	3
Physical appearance			
Colour.	Normal	Normal	Normal
Clearness.	Clear	Clear	Clear
Smell.	Nil	Nil	Nil
Taste.	Nil	Nil	Nil
Sediment on ignition.	No changing	No changing	No changing
Chemical determination (parts per 100,000)			
Total solids.	9	5	5
Chlorine.	0.56	0.56	0.56
Ammonia free.	Nil	Nil	Nil
Albuminoid Ammonia.	Trace	0.005	0.005
Nitrites.	Trace	Nil	Nil
Nitrates.	Nil	Nil	Nil
Iron.	Nil	Trace	Nil
Oxygen absorbed in 15 minutes.	Trace	Trace	Trace
Flourine.	Nil	—	--
Total hardness (Grains per gallon or parts per 70,000).			
Total hardness.	2.8	3.5	4.55
Date of analyses.	17th Jan. 1940.	11th Aug. 1939.	19th Dec. 1939.

These data tend to reveal effectively that alluvial deposits formed by different rivers would not only have high amounts of mineral matters, but they would also vary in composition. Therefore, all alluvial soils will not be uniformly productive for one and the same types of vegetation. For example, river alluvia which have high amounts of Ca salts may have a community of *Acacia catechu—Dalbergia sissoo,* while another with sodium salts may have *Tamarix dioca* as the dominant plant. In the Himalayan valleys, neutral sandy river alluvium bears *chir* pine and blue pine, while those with high amounts of Ca usually have *Ulmus wallichiana, Alnus nitida, Populus ciliata,* etc. In the Dun Valley, some neutral alluvia bear grasses and no tree vegetation.

The alluvial deposits derived from trap country in the Deccan can support good teak plantations, while those derived from other rocks have bamboo as the chief plants. The analysis of river water as well as alluvial soils formed in different climatic regions would therefore provide useful data for raising plantations.

The data presented also show the seasonal differences in the quantity of mineral matter dissolved in water that may be related to temperature changes.

The analysis of waters from some well known subterranean waters of India are given in Table 36. This also shows that

TABLE 36

The composition of three well known subterranean waters in India, as parts per 100,000.

(After N.C. Bose, 1940)

	Spring near Monghyr	Spring near Hazari-bagh	Tubewell Victoria Hotel Puri
Dates on which analysed	10-7-1929	20-1-1930	19-4-1933
Calcium carbonate.	8.33	0.63	26.0
Magnesium carbonate.	1.73	trace	2.11
Sodium carbonate.	0.32	8.88	
Sodium sulphate.	—	8.6	—
Sodium chloride.	4.46	14.03	29.78
Sodium nitrate.	0.53	—	10.8
Silica.	—	10.22	—
Others.	3.36	10.48	5.2
Magnesium sulphate.	—	—	2.4
Magnesium chloride.	—	—	2.35
Total solids.	18.73	52.84	78.64

there is a variation in the amount of dissolved mineral salts in water from different regions.

The composition of water of some Punjab rivers is given in Table 37.

TABLE 37

Composition and concentration of salts in the waters of Punjab rivers and canals.

(After Asghar and Dhawan, 1948).

Sampling site.	Salt content (Parts per 100,000) Calcium.		Sodium			Total	Conductivity.
	Sulphate	Bicarbonate	Sulphate	Bicarbonate	Chloride		
Western Jumna Canal near Abdullapur.	4.8	10.0	1.9	—	1.8	18.5	260
River Sutlej near Phillaur.	3.4	16.0	3.2	—	2.3	24.9	325
Sirhind Canal near Doraha.	4.4	11.0	3.9	—	2.3	21.6	325
River Beas near Dhilwan.	7.5	11.0	—	—	3.5	22.0	260
Upper Chenab Canal near Chichoki Mallian (U.C.C.).	—	10.0	11.4	1.1	2.9	25.4	275
Upper Bari Doab Canal near Tibri (U.B.D.C.).	3.4	7.0	3.6	—	1.8	15.8	200
Lower Gungera branch Lower Chenab Canal near Jaranwala (L.G.B.).	8.8	11.5	1.4	—	2.3	24.0	250
Lower Jhelum Canal near Ludewala (L.J.C.).	3.4	15.0	3.6	—	4.7	26.7	340

The data show that the amount of salts vary a good deal in different rivers, therefore, the productivity of alluvia of these will also vary accordingly.

Some analysis of Calcutta filtered water given by Coulson are presented in Table 38.

TABLE 38

Calcutta filtered water during 1938.

Results of analysis expressed in parts per 100,000.

(After Coulson, 1940).

Month.	Total solids in			Suspended solid or silt in		Chlorine.	Saline ammonia	Albuminoid ammonia	Total hardness	Permanent hardness	Nitrogen as nitrites and nitrates	Oxygen consumed in two hours	Nitrites	pH value	Alkalinity
	Raw water	Settled water.	Filtered water	River water	Settled water										
January	37.3	25.08	24.2	10.3	2.08	1.26	0.00092	0.00384	12.36	5.88	0.0108	0.01278	Nil	7.5	25.64
February	46.45	28.7	24.4	16.7	1.4	1.42	0.0009	0 0036	13.25	6.27	0.011	0.0128	Nil	7.5	24.1
March	46.2	25.7	27.6	15.1	1.2	2.75	0.00095	0.0042	12.5	5.90	0.011	0.0121	Nil	7.5	23.4
April	49.5	26.5	23.6	27.7	1.75	4.5	0.001	0.0043	13.1	6.80	0.117	0.01397	Nil	7.5	21.9
May	61.9	28.3	25.7	35.4	2.9	6.3	0.001	0.00432	13.4	6.88	0 0124	0.01132	Nil	7.52	20 5
June	106.0	26.9	23.07	73.3	2.3	1.83	0.009	0 0045	11.8	5.70	0.015	0.0145	Nil	7.5	11.3
July	89.6	22.8	19.4	60.6	2.07	1.02	0.0011	0.0045	10.9	5.5	0.0135	0.0131	Nil	7.5	8.5
August	51.2	18.16	15.1	27.1	1.7	0.75	0.00116	0.00464	9.84	4.84	0.0136	0.1288	Nil	7.4	8.2
Sept.	35.2	15.3	12.3	16.0	1.3	0.80	0.00115	0.005	9.2	4.28	0.0115	0.01302	Nil	7.4	8.07
October	31.28	14.14	12.0	15.9	1.2	0 83	0.00108	0.00432	9.22	4.08	0.0134	0.01294	Nil	7.4	8.96
November	25.4	15.4	13.8	5.87	1.2	0.837	0.001	0.00285	9.87	4.65	0.0117	0.012127	Nil	7.45	11.9
December	37.87	18 57	15.7	18.5	2.02	1.012	0.00105	0.005	10.65	5.2	0.0125	0.01222	Nil	7.5	17.0

Raw water contains greater amounts of total solids and on settling some of these are got rid of. Much of suspended substances settle down soon. The seasonal variations in soluble salts are also shown in this table.

The beneficial effect on vegetation of subterranean waters which have large quantities of dissolved substances may thus be clear. In regions of low rainfall subterranean waters will have different effects on plant growth than in regions of higher rainfall.

Solutes from atmosphere—The composition of the soil water is also influenced by the content of dissolved gases from the atmosphere. Lightning flashes during rains oxidise atmospheric nitrogen and yield nitric acid. The atmospheric oxygen, nitrogen. carbon dioxide, sulphur oxide, chlorine, etc. which are dissolved by falling rain drops are added to the soil. The effect of carbonated rain water in dissolving Ca salts from the soils is well known.

In Tables 39, 40, 41, 42 the amount of dissolved oxygen, chlorine nitrate, nitrogen and ammoncial nitrogen added to the soil in different places in rain is presented.

TABLE 39

Dissolved oxygen brought down in rain.

(After Richards, 1917).

	Average rainfall (in.) at Rothamsted (28 years).	Dissolved oxygen parts per million	lb. per acre
Summer.	13.32	9.0	27.12
Winter.	15.50	11.2	39.27
Year.	28.82	—	66.39

TABLE 40

Dissolved chlorine brought down in rain.

(After Miller, 1913).

	Annual rainfall in.	Chlorine per acre lb.
Laudale.	76.89	168.5
Butt of Lewis.	40.57	6884.0
Monarch.	47.21	2723.0
Barrahead.	33.93	5753.0

TABLE 41

Nitrogen in the rainwater at Ithaca, New York between May 1, 1915 and May 1, 1920.

(After Wilson, 1921).

Month collected.	Rainfall inches.	N as NH_3 per acre lbs.	N as NO_3 per acre lbs.	Total N per acre lbs.
May	19.89	9.18	0.82	10.00
June	18.11	10.37	0.51	10 88
July	19.38	15.00	0.40	15.40
August	18.10	5.59	0.67	6.26
September	17.61	3.70	None	3.70
October	15.26	2.11	0.26	2.37
November	5.98	0.82	0.27	1.09
December	5.06	2.66	0.24	2.90
January	3.68	0.99	0.29	1.28
February	4.96	1.11	0.43	1.54
March	7.00	1.94	0.30	2.24
April	11.51	4.06	0.85	4.91
Total	146.54	57.53	5.04	52.57
Yearly average	29.31	11.50	1.01	12.51

TABLE 42

Amounts of nitrogen brought down in rain.

(After Lyon and Buckman, 1937).

Location.	Years of record number.	Rainfall (inches).	Nitrogen added per acre per year. Ammonical nitrogen (Pounds).	Nitrogen added per acre per year. Nitrate nitrogen (Pounds).
Harpenden, England.	28	28.8	2.64	1.33
Garford, England.	3	26.9	6.43	1.93
Flahult, Sweden.	1	32.5	3.32	1.30
Groningen, Netherlands.	—	27.6	4.54	1.54
Bloemfontein and Durban, South Africa.	2	—	4.02	1.39
Ottawa, Canada.	10	23.4	4.42	2.16
Ithaca, N. Y.	11	29.5	7.10	0.80

In addition to oxygen, chlorine and nitrogen, sulphur is also brought down in rain as sulphate. Lyon and Buckman (1947) state that 2-3 pounds of sulphate an acre per year may be added to soils; and if the area is near the industrial centre the amount per acre may be as high as 100 pounds. As shown in Table 43 the amount of sulphur brought down in rain at the Iowa Agricultural Experimental Station differs during different months of the year.

TABLE 43

Content of sulphur brought in rain during 1921 at Iowa Agricultural Experimental Station, U.S.A.

(After Erdman, 1922).

Month.	Rainfall (inches).	Sulphur content.	
		SO_3 Lbs. per acre.	S Lbs. per acre.
January and February	1.0	5.04	2.02
March	1.28	3.75	1.50
April	2.91	5.49	2.20
May	2.83	4.57	1.83
June	4.77	2.67	1.07
July	1.11	1.00	0.40
August	7.35	5.47	2.19
September	6.76	3.77	1.51
October	1.25	3.67	1.47
November and December	1.12	1.75	0.70
Total ...	30.38	37.17	14.89

Solutes from minerals. When rain water charged with CO_2 reaches the subsoil as underground water, it acts on various soil minerals and brings them into solution. The chief rocks that are affected by this are limestones, dolomites and other calcium containing formations. By the dissolution of calcium from rocks the soils left are only composed of the impurities, that were originally present in such rocks. The effect of rain on limestone rock will be clearly seen in Burma,

Andamans and other areas where rainfall is usually higher than 80 inches per year.

The clay-with-flints soils capping the Jurassic Chalk in southern England are formed by the dissolution of calcium carbonate from the parent rock and are thus non-calcareous and often base deficient.

The physical effects of this process are seen in the sinking of the entire bodies of rocks and the formations of basins or funnel-shaped hollows, which are characteristic of limestone regions. Small caves and underground channels known as caverns in limestone areas are well known. Sometimes whole valleys may be carved in the limestone country by rain waters. The formation of calcareous tuffa and calcareous marl in the plains of India along river courses is due to the dissolution of limestone rocks by carbonated rain waters and their re-deposition at lower levels.

Stalactites and stalagmites are formed when water is evaporated from calcareous solutions. Concretions and kankar, so well known as a pan layer, in the Indo-Gangetic alluvium had their origin in the same way. The dissolution of calcium rocks and consequent formation of valleys, hollows, sinks, caverns, etc. with their characteristic rugged and dry appearance is distinctive of limestone regions and such a topography is called *karst* topography.

Underground waters occur comparatively near the surface of the soil, if vegetation is kept intact. In a synclinal basin a bore in the middle of the valley may produce artesian wells, and the maintenance of these is very greatly influenced by the vegetal cover in the area, especially on the escarpment of the basin. A dense vegetation cover tends to keep the ground water nearer the surface, and small rivulets and streams issuing from small valleys remain active for a greater part of the year. But once the vegetation is removed the spring-heads get dry and valley channels no longer retain any water. This effect of vegetation on ground water has been shown in the London Basin, where in the Downs during high rainfall years dry springs in valleys become active for a short time by temporary rise of underground water.

Base exchange and conservation of plant nutrients.—The solutes in the soil water are delicately balanced in a complex system of weathering mineral and organic particles, colloidal complex solution and plant root. As they are released from the weathering material into the solution much of the cations undergo base exchange. The micellae of the clay colloids are negatively charged and hold cations of NH_4, K, Na, Ca, Mg, Fe, etc. adsorbed on their surface. These are exchanged for other cations which may happen to be in higher concentration in the surrounding solution. Thus, when a soil is shaken in acid solution H^+ replaces the other cations in proportion to its concentration. A sodium salt solution will similarly remove the cations and enrich it with Na^+ and so on. In the base exchange phenomenon monovalent as H^+, NH_4^+, K^+ and divalent as, Ca^{++} are more reactive than the trivalent cations, Al^{+++} and Fe^{+++}.

As the plant absorbs a particular ion the soil solution gets replenished in respect of it from the colloidal complex by base exchange. The ion supplied to the colloids in exchange of those absorbed by the plant are released either from the plant itself by respiring roots in form of H^+ from H_2CO_3 or other cations from the weathering complex. It is also thought that the root hairs might be in the same fashion absorbing ions by contact with the micellae.

The base exchange phenomenon in nature is of more significance than has been hitherto considered. Drainage water removes all the soluble salts from a soil within a short time, creating famine conditions for the plant, were it not for the soil colloids to conserve them by adsorption. Furthermore, any solute released in a higher concentration from the weathering complex would make the solution toxic to plants if the same is not partly removed as is done by the soil colloids. Nitrates are not adsorbed and hence drained soils become deficient of these on rainfall. However, their production in such soils is a continuous phenomenon as has been already explained.

9

Soil Air System

Howsoever the soil particles be closely packed, there will always remain spaces between them on account of their configuration. There are pore spaces which are interconnected by their counterparts of smaller or larger size. They are occupied by the soil solution and gases in reverse proportion. The latter constitute the soil air which is continuous with the atmosphere above the surface. The gases present in both are just the same but their proportion is different. Under all situations the concentration of oxygen is lower and that of carbon dioxide higher in the soil air so that a steep gradient for diffusion between the two exists. Since the growth of roots, germination of seeds and micro-biological activities within the soil depend upon the extent of aeration the amount and quality of the soil atmosphere is of great importance to plant life.

Pore space.—Larger particles on packing will collectively leave smaller amount of pore space than finer particles. Light soils and heavy soils range between 40-60 per cent of pore space on the average. However, the size of the pore space will differ in the two cases. Sand has larger spaces through which gravitational water flows down and clay has much smaller pores which holds the water by capillarity. Hence the pores of the former are called non-capillary and those of the latter capillary pores. The non-capillary pores are all the time filled with air, except when the soil is waterlogged. The capillary pores on the other hand contain fluctuating amounts of air depending upon the movement of the soil solution.

Ordinarily soils contain both capillary and non-capillary pores according to their structure. In heavy soils the clay fraction is dispersed and the pore spaces are very small which get choked up on wetting. However, lime, roots, organisms, humus and gum-like substances aggregate these particles leaving larger pore spaces for soil aeration. The soils become 'lighter' and 'open' when so treated and develop 'crumb' structure in which both water and air are retained in suitable proportions. Root channels and animal burrows besides good aggregation, increase the porosity of soils.

Composition of soil air.—The gases within the pores move by diffusion according to their individual concentration gradients. The roots and the living organisms are depleting oxygen and releasing carbon dioxide all the time-more so at higher temperatures. As soon as the concentration of oxygen falls lower than 10% anaerobic conditions are set up. The carbon dioxide concentration may rise to more than 1%, sometimes even 10%. Both the conditions are toxic and stop root growth and biological activities of aerobic organisms. However, downward diffusion of oxygen from and upward diffusion of carbon dioxide to the atmosphere, restore equilibrium. Since diffusion is a slow process and soil respiration is higher in magnitude the soil air will always be poor in oxygen and rich in carbon dioxide, more so in deeper layers. Roots, therefore, are largely confined to the surface layer of the soil.

Movement of the gases.—Although, aeration is inversely related to soil moisture the movement of water on periodic wetting and desiccation of the soil improves gaseous conditions in it. During alternate heavy showers and drought of the monsoon rains the non-capillary pore spaces get filled up with water and air in quick succession. The air is expelled to the atmosphere and sucked in again from there as the moving head of gravitational water goes down. Thus, there is a mass mixing up of the soil and the atmospheric air bringing about better aeration even in deeper layers.

Aeration and plant growth.—Poor soil aeration leads to suppression of root hair and root development. Absorption of water and nutrients is also reduced. However, the roots of aquatic plants and those of water-logged areas seem to suffer little from these effects as they are adopted to such conditions by storing air in their aerenchyma system throughout the body.

Carbon dioxide is readily soluble in water. As such the gas when in higher concentration gets dissolved in it and lowers the pH value thus setting up secondary effects on plant growth. Whitney and Gardner's (1943) data on the effect of CO_2 on pH of soil suspensions are given in Table 44. These show that significant variations in the pH values of soil are obtained by the effect of CO_2.

TABLE 44

Effect of CO_2 on pH of soil suspensions.

(From Whitney and Gardner, 1943)

Soil No.	1 soil to 2 water		1 soil to 5 water		1 soil to 10 water	
	Atmospheres	pH	Atmospheres	pH	Atmospheres	pH
1	.00025	9.03	.00025	9.00	.00025	8.90
	.00042	8.93	.00042	8.85	.00042	8.63
	.09687	7.06	—	—	—	—
	.77893	6.40	.7826	6.38	.76826	6.28
2	.00034	8.47	.00037	8.08	.00039	7.92
	.02914	7.01	—	—	—	—
	.77893	6.08	.76679	5.94	.76751	5.88
3	.00034	6.88	—	—	—	—
	.02914	6.28	—	—	—	—
	.77893	5.70	—	—	—	—
4	.00034	8.06	.00037	7.29	.00039	7.21
	.09687	6.38	—	—	—	—
	.77893	5.71	.76679	5.37	.76751	5.08
6	.00070	7.88	—	—	—	—
	.09687	6.70	—	—	—	—
	.77893	6.13	—	—	—	—
8	.00030	9.20	.00030	9.03	.00030	8.90
	.00062	8.96	—	—	—	—
	.05372	7.45	—	—	—	—
14	.00029	7.50	—	—	—	—
	.02453	6.41	—	—	—	—
	.77278	5.69	—	—	—	—
15	.00029	7.07	—	—	—	—
	.02453	6.35	—	—	—	—
	.77278	5.78	—	—	—	—

In India, the deleterious effect of CO_2 on plant growth has not been studied much, however, some observations were made by Griffith and Gupta (1947) in *sal* forests of Dehra Dun. These extended earlier observations of Hole (1911) who first showed that the cause of mortality in *Shorea robusta* seedlings in nature was probably the result of high concentration of CO_2 in the soil atmosphere and soil solution.

It may be interesting to refer to the important works of Rommel on soil aeration of forest soils and of Russell on agricultural soils of England.

10

Development of Soils

Factors : From the ecological standpoint the importance of the soil factor lies in the fact that it provides medium for plant growth. There may be five types of environmental factors that govern the development of soil in relation to vegetation development. First of these is the *parent material,* the nature of which has been already discussed. The second set of factors are *climatic,* which have a pronounced effect on the processes of weathering and disintegration of rocks into soils. Of these the amount of rain water percolating through the soil body in any season of the year, which process is called leaching, is of great importance. Leaching removes from the soil body soluble substances like lime and other bases and plant nutrients, specially nitrates.

Leaching and loss of soluble substances are determined by lateral and vertical movement of water down a slope, therefore, *relief* assumes importance in the development of soils.

The effect of all these factors is not instantaneous, therefore, a *time* factor has also to be considered. *Vegetation* exercises pronounced effects on the development of soil through its ability to extract minerals from lower layers of soil and bring them on the surface in the litter, annually. The fertility of soil is the result of this biological circulation of plant foods and naturally those plants which have greater demands on the soil will produce fertile type of surface soil. This also acts against surface leaching and in regions with low leaching, surface soil usually becomes more fertile with the progress and development of vegetation.

The realisation of the non-exclusive importance of the climatic factor in soil development was made by Russian workers and it was later on recognised in other countries. The Russian influence on pedology is shown by the terms podzol, cernozem, krasnozem, solonets and terra rossa, etc. which were first defined in Russia and were now adopted universally.

The consideration of the importance of the factors of the soil development indicated above has led to the understanding of local soil differences and it has also enabled the formulation of the following three laws of soil distribution on world basis :—

(a) The law of horizontal (latitudinal) zonality;

(b) The law of provinciality, or regionality;

(c) The law of vertical zonality.

In the classification of the great soil groups of the world shown in Fig. 25 and those of the United States by Marbut and Kellog, the effect of such factors has been considered.

The exclusive role of climate in determining the distribution of soil types has been attempted recently in Australia by Prescot who was able to show a close correlation between a certain number of soil groups with the rainfall evaporation index $P/E^{0.7}$. Reference to this work and that of Thornthwaite in this connection will be made in the section on climate.

Ecological Classification of Soils

For the growth of plants, soils are usually classified into the following categories (1) Skeletal and Immature soils (2) Flushed and enriched soils (3) Leached and impoverished soils (4) Water logged soils.

1. **Skeletal and Immature soils** represent early stages of soil development in which rock fragments are predominant. They occur in regions of low temperatures, or where the underlying rocks are hard, and by these factors the soils tend to remain unstable and do not show much development. All mountain soils and freshly deposited soils in landslips, slides, or surface creeps, moraines, alluvia, etc. come under this category. They truly represent dynamic systems. The effect of low temperature at high altitudes in the Himalayas is seen in the

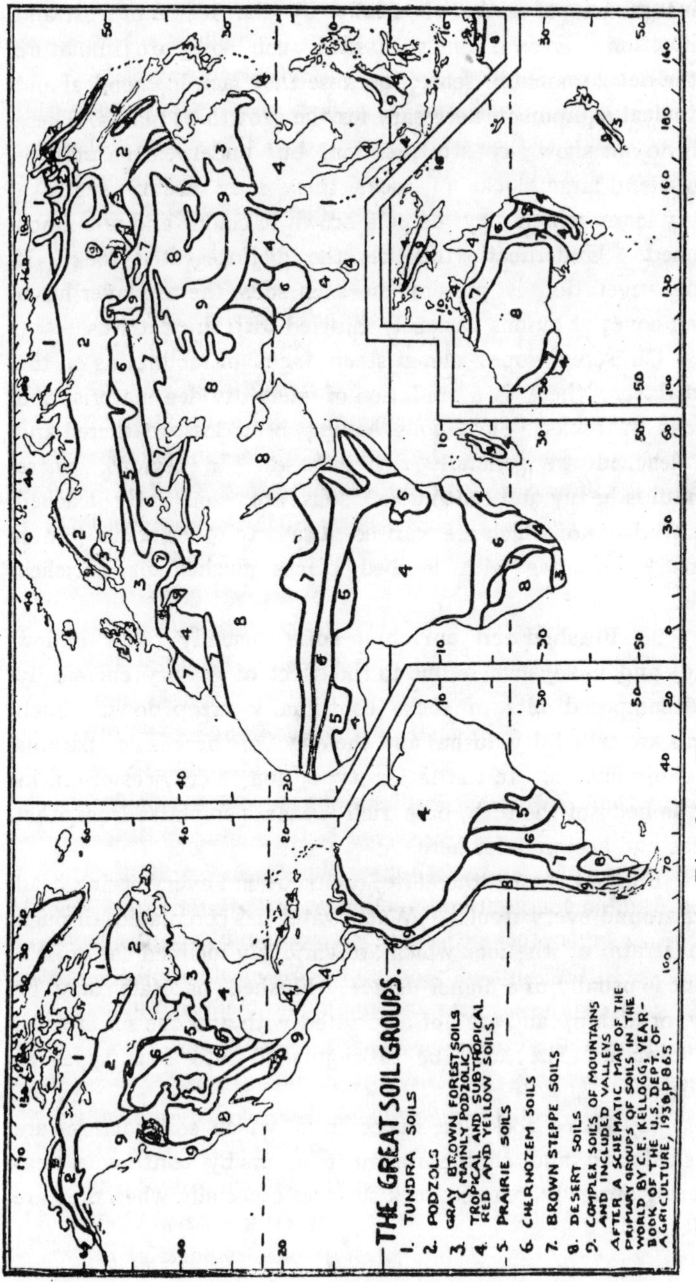

Fig. 25. The great soil groups of the world.

disintegration of rocks physically by the action of frost and solifluction. Even if finely divided such soils are immature in the developmental sense, because they lack in chemical and biological equipment, necessary for the growth of plants. These soils do not show any stratification, but underneath a mass of coarse and large blocks of rocks there may occur usually a sandy loam, which may be pale brown in colour and not much leached. Underneath this lies the distintegrated rock. If some vegetation is present on such soils, the top layer has a black cover of humus variously mingled with the detritus.

On scree slopes along steep faces of mountains on the Himalayas, there is a gradation of finely divided material and blocks of rocks. Such soils having been less weathered and not leached are generally rich in bases. In regions, where rainfall is heavy and continuous, soils are constantly leached the products of which are carried away to other areas. Such areas have along with leached places flushed or enriched soils.

2. **Flushed and enriched soils** usually occur on low lands and valleys where due to the effect of gravity chemically undecomposed bits of rocks continually creep down. Such areas are called dry flushes and the effect of these is comparable to damp flushes. In the latter type due to crevices or cracks in the body of the rock, base rich water percolates from other areas and accumulates here.

Flushed areas, therefore, occur around every spring head and around every rivulet. When water has percolated, through the strata of the rock which are variously inclined the enrichment is usually of a higher degree. Flushed habitats may be dry, or damp, and are not associated with any one set of soils, or climate. They may be distinguished into four types as under :—

(a) Bare rock, or slopes from which soil particles are washed away and thus enrichment occurs by continuous weathering of newly exposed rock surfaces especially when they are soft.

(b) Block scree in which, although, leaching may be more than weathring in humid climate, the distintegration replenishes some of the bases lost from the soil, especially if the rocks contain high quantities of lime. In arid areas block screes may be richer in bases, but definite information for this under tropical conditions is at present lacking.

(c) Unstable scree slopes and solifluction areas in which by movement and accumulation of soil particles enrichment occurs by weathering and particle accumulation.

(d) Accumulation areas occur where flushes from any of the above types operate. In such areas the soils are usually deep, fine and fertile.

Base rich water may occur in any of these four types of flushed soils and, although, it is the rarest in the type (a) and commonest in (c) and (d), the flush effect may counteract leaching losses, if the rocks are lime-rich.

As leaching is intense due to humid climate in northern latitudes, the effect of flushes on the development of vegetation is distinctly seen and well understood in those areas, and especially if the underlying rocks are poor in lime. In the Himalayas, however, as leaching is at no place so severe as to produce base poor soil the effect of flushing and its role in vegetational development is not so clearly appreciated. The occurence of better and greener types of vegetation round a rivulet, or a springhead is usually considered to be due to the effect of comparatively large amount of moisture alone and the role of minerals dissolved in the water, which have really produced the beneficial effect on plants is not fully appreciated.

On mountains of the plateau region, where ancient rocks occur the effect of flushes on plant growth may be better appreciated, but no study of this type has so far been made in this country.

(3) **Leached and impoverished soils** occur in northern latitudes where climate is humid and precipitation is always in excess of evaporation. When unstable surfaces become stabilised and have a cover of vegetation they inevitably tend to become leached to various degrees, depending upon

the porosity of the soil, topography and type of vegetal cover. This type of development is especially seen in mountains of Europe. Under the influence of humid climate, bases are constantly leached and a layer of undecomposed plant material accumulates on the surface and the soil exhibits a characteristic profile indicating that it has reached its development.

A hypothetical soil profile is shown in Fig. 26. It shows two main layers, *alluvial* and *illuvial*. The former is also called leached horizon, from which bases are lost and deposited in the lower illuvial layer, sometimes forming humus or iron pan. In the extreme form of development, which is called podsolization there develops a peaty humus layer on the surface of the soil from which all available bases, especially lime and

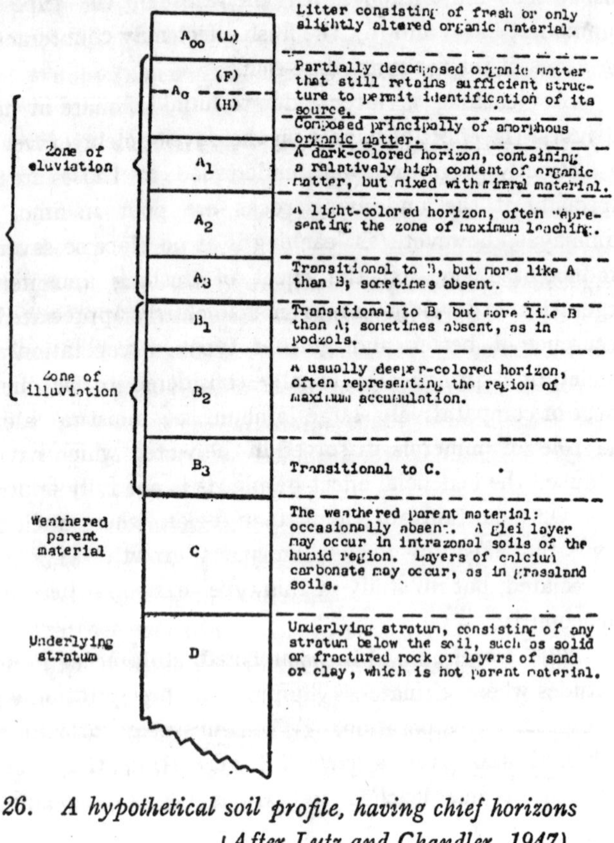

Fig. 26. A hypothetical soil profile, having chief horizons (After Lutz and Chandler, 1947).

iron are leached away. The podsol profile usually has this type of surface layer below which pan layers are recognised.

Podsols are characteristic of the cold temperate regions, having coniferous vegetation ; and they may be found in regions of moderate rainfall, if the underlying rocks are stable, lime poor and porous.

In India, truly podsol soils have not been discovered even in Silver fir forests at an altitude of 10,000 feet, or above. This is on account of the fact, that the climate is not uniformly and continuously humid, but periods of intense precipitation always alternate with periods of high evaporation ; as a result of which base-rich water rises up the profile through capillary action and tends to make up for the loss of the bases which have been leached down through excessive percolation of rain water. The characters and statification of various chemical properties in the Himalayan podsols are compared with northern podsols in Fig. 27. Profiles 57 and 40 on the left hand side in the figure are from the Himalayas and the three profiles on the right are from Sweden. It follows that the upper layers of the soils which should be acidic and base deficient as in a normal podsol profile are not so in the Indian region.

Between the extreme podsols and immature soils there occur a range of soil profiles, usually classed under the category of brown earths, on account of their having ochreous brown colour. In Europe, these occur in the deciduous forests of broad-leaved species, usually in low lands. In these the surface layers are only somewhat leached of bases, retaining brown colour and sufficient amount of bases to produce a moderate soil fertility.

In the Himalayas, most of our forest soils are of the brown earth type. Here again, these are different from the European brown earths and salient properties of the coniferous brown earth profiles of the Kulu Valley have already been compared with European brown earths.

Depending upon the amount of bases which the litter of tree vegetation releases to counteract the process of leaching brown earths in Europe exhibit a great variety of types. If the brown earth soils or gray-brown podzolic soils are deforested

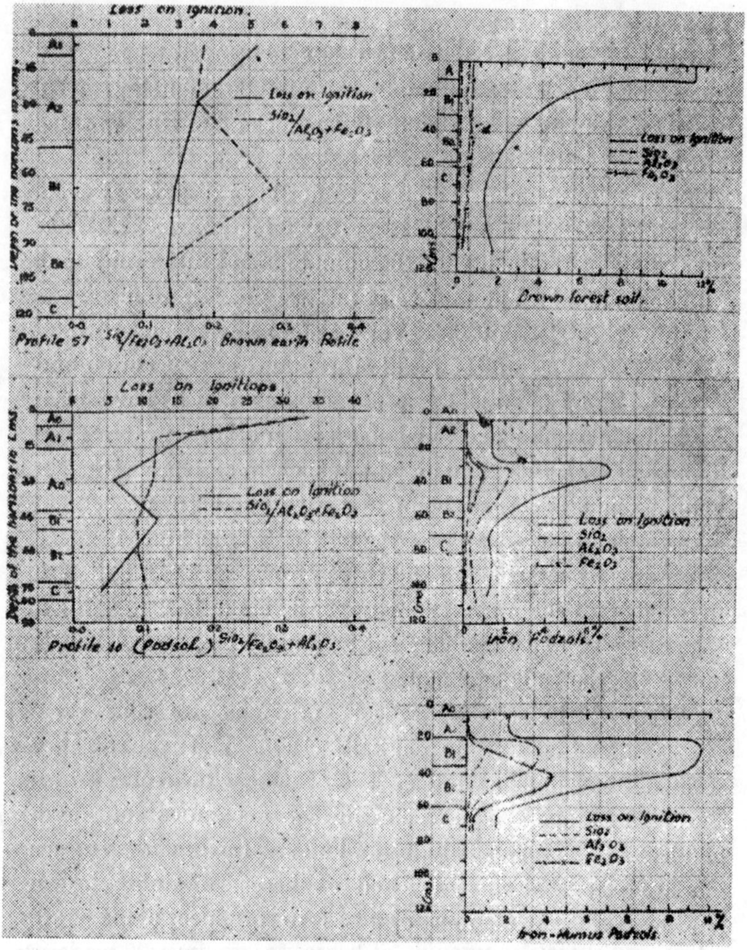

Fig. 27. Comparision of chemical properties of two Himalayan soils (left) with some typical Swedish soils (right).

the accumulated humus is rapidly oxidised and the upper layer of the soils are lost by erosion and rainwash and the lower horizon of the soils is exposed on the surface. Such profiles are called *truncated profiles*, from which the top strata have been removed ; and these have a high base status like immature soils (Fig. 28).

Under intense climatic conditions if the soil is porous, the development of the podsol reaches its maximum and an impermeable pan layer which impedes drainage is found below the soil (Fig. 29). This results in water-logged conditions especially if the rainfall is great and leads to the accumulation of peat. Once the peat accumulation sets in leaching is greatly accentuated due to poor drainage and organic acids produced by peats. Thus, at high altitudes in the northern regions of Europe, podsols have almost always tended to become peat covered bog soils. Due to the excess of moisture in the profile the characteristic horizons of a podsol disappear and destroy its

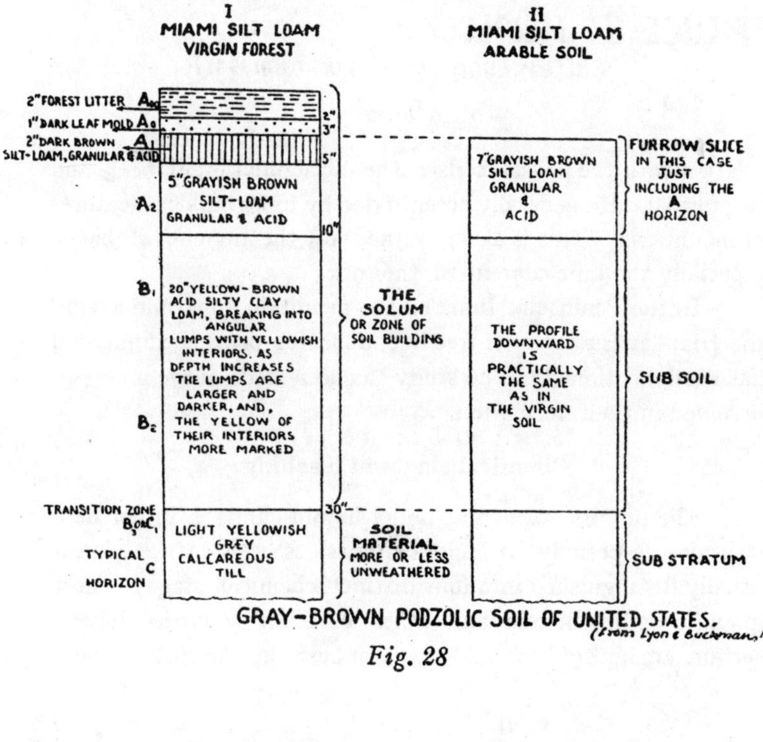

GRAY-BROWN PODZOLIC SOIL OF UNITED STATES.
(From Lyon & Buckman, 1947.)

Fig. 28

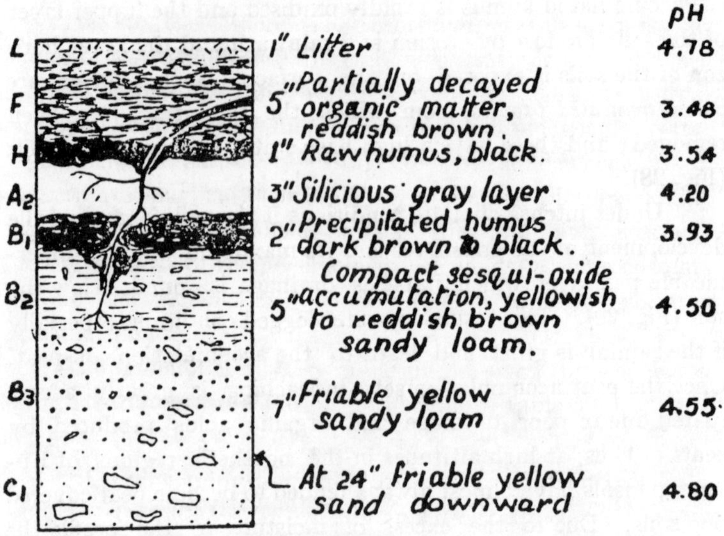

MATURE PODSOL, NEW YORK UNDER BIRCH, SPRUCE & HEMLOCK.
(after Lyon and Buckman, 1941)

Fig. 29

relation with the podsol soils. The accumulation of peat and bog formation is generally accentuated by topographical features of mountains. This is also governed by the amount of bases, especially the lime content of the rock.

In the Peninsular India on the mountain tops and also in the Himalayas above the tree line, doubtful peat accumulation has been mentioned but no study has been made of the processes involved in their formation.

Chemical stages in leaching

The process of leaching under humid climates in northern latitudes, (especially at high altitudes) is well marked and is usually distinguished into four distinct chemical stages. Most upland soils in Britain and other northern countries have a certain amount of lime as the main base in the rock, These soils initially have a high fertility but during the process of leaching, lime and iron salts are removed and their place is taken up by hydrogen, as a result of which the soil ultimately becomes acid and sour to the taste. On account of oxidation processes in the soil, iron is not leached to any great extent, therefore, the colour of the soil is brown. According to the degree of leaching the following four stages represented by different types of soils in northern countries can be usefully recognised :—

(1) Lime saturated: pH of these soils is usually above 6.5 and they effervesce with dilute Hcl.

(2) Lime deficient: These have a pH between 5 and 6 and do not show any effervescence with dilute Hcl.

(3) Base deficient: These have a pH between 4 and 5 at which iron becomes soluble and relatively more important than lime. When treated with Comber's reagent they produce a red colour.

(4) Acid soils: These have pH below 3.8 and are characteristically acid soils, usually associated with podsolic profile and are often marked by peat accumulation, or heath formation. When shaken with Comber's reagent they produce a deep red colour.

The stages two and three are usually referred as brown earth soils and their recognition is of great ecological importance, for a well marked transition in vegetation occurs at a point between 2 and 3, which is approximately a point of half saturation with bases.

These four soil types can be distinguished by their characteristic appearance and ecological properties. The grey colour in a podsol profile indicates it to be mainly a hydrogen soil; and orange brown colour may indicate that most of the lime has been removed and iron has become mobile and oxidised in the presence of oxygen. Red and blue spots in a soil profile may indicate the condition of water-logging.

In the process of leaching and soil development, increased acidity resulting from formation of oxalic and citric acids, in which iron salts are especially soluble, and the establishment of the water-logged conditions, which by excluding oxygen and allowing the formation of ferrous iron that is much more soluble, accentuate podsolisation. The colour of the soil, therefore, gives a fair indication to the field ecologist of the progress of leaching and biological properties of the soil.

The conditions sketched above refer to humid climates and it is doubtful, whether any such processes occur in this country even in so called temperate regions of India, exactly in the same manner as in northern countries. A detailed study, however, is very desirable.

Profiles of Some Indian Soils

The soils of India are essentially mineral soils with very little humus and without a characteristic profile development. A number of studies have however, been made on the vertical stratification of minerals by digging pits or removing columns of soil. These are called profiles for the sake of convenience rather than on developmental basis. A few profile studies are given in the following Tables 45 and 46.

Profile figures of some soils bearing open deciduous forests with teak and *Terminalias* are given in figure 30.

In the Himalayas, there is some development of horizons in soils. Weakly podsolised and brown-earth soils resembling European soils have been recognised under different communities of conifers (Taylor, Mahendru, Hoon and Mehta, 1938; Puri and Gupta, 1951). Some profiles studied by Puri from the Bashahar Himalayas (see Mohan and Puri, 1956) are given in figures 31 and 32.

Two soil profiles from the sal forests of the Dehra Dun Valley are shown in Fig. 33. The distribution of organic matter content and exchangeable Ca is also shown in those profiles. They show very interesting mode of development, which was shown by Puri (1950) to influence greatly the distribution and succession of forest communities under natural conditions.

TABLE 45

Profile distribution of soil characteristics expressed in percentage of dry weight of soil for the forests of Upper Godavari Division

(After J. Banerji, 1941).

Vegetation	Pit 2 Chestnut earth "lanka" Teak forest good quality.			Pit 1 Truncated chestnut earth "Sundra Boru" Teak forest poor quality or none.	
	A layer 1½" from top	B layer 1' from top	C layer 2' from top	A layer 1½" depth	B layer 10½" from top
Organic matter. ...	2.36	1.90	1.65	2.79	2.75
Moisture (on air dry soil). ...	2.75	3.36	2.14	1.26	1.88
Percent calcium carbonate. ...	0.13	2.59	1.48	0.04	0.05
Total nitrogen. ...	0.026	0.012	0.008	0.067	0.035
Available P_2O_5. ...	0.002	0.002	0.002	0.025	0.007
Available K_2O. ...	0.007	0.006	0.005	0.011	0.007
SiO_2/R_2O_3 Ratio, in clay (molecular ratios used)...	2.61	2.73	3.15	2.47	2.50
pH values (1.5 suspension). ...	7.1	8.0	8.0	6.7	6.4
Total exchangeable bases (milli-equivalents). ...	13.20	16.40	11.14	6.39	9.29
Exchangeable Ca. (milli-equivalents). ...	7.01	6.40	4.03	5.28	5.33
,, K and Na (,,). ...	2.22	5.90	3.71	0.31	0.31
,, Mg. (,,). ...	3.97	4.10	3.40	0.80	0.65

TABLE 46

Composition of two profiles of trap hill soil bearing open deciduous forest.

(After Bhatia, 1955).

Depth.	Pit 1					Pit 2			
	Surface	6″	12″	20″	32″	Surface	6″	12″	20″
pH	6.85	6.80	6.85	6.85	8.05	6.60	6.45	6.30	6.10
SiO_2	49.68	47.98	47.68	46.14	50.85	53.26	52.23	51 01	47.59
R_2O_3	0.70	30.85	32.40	33.52	31.18	27.65	30.12	29.45	33.45
$\frac{SiO_2}{R_2O_3}$	1.62	1.55	1.57	1.38	1.63	1 93	1.73	1.73	1.39
CaO	1.25	1.29	1.90	1.80	2.57	0.72	0.86	1.00	1.02
$Mg_2P_2O_7$	1.85	2.40	3.36	1.52	4.85	0.95	1.20	1.50	1.52
Available R_2O_3	0.07	0.09	0.1	0.11	0.13	0.08	0.09	0.09	0.10
Available CaO	0.53	0.52	0.53	0.50	1.95	0.35	0.35	0.34	0.31
Available $Mg_2P_2O_7$	0.3	0.33	0.35	0.22	—	0.39	0.32	0.31	0.25
Org. matter	3.96	2.85	2.04	1.14	—	3.78	2.48	2.44	1.38

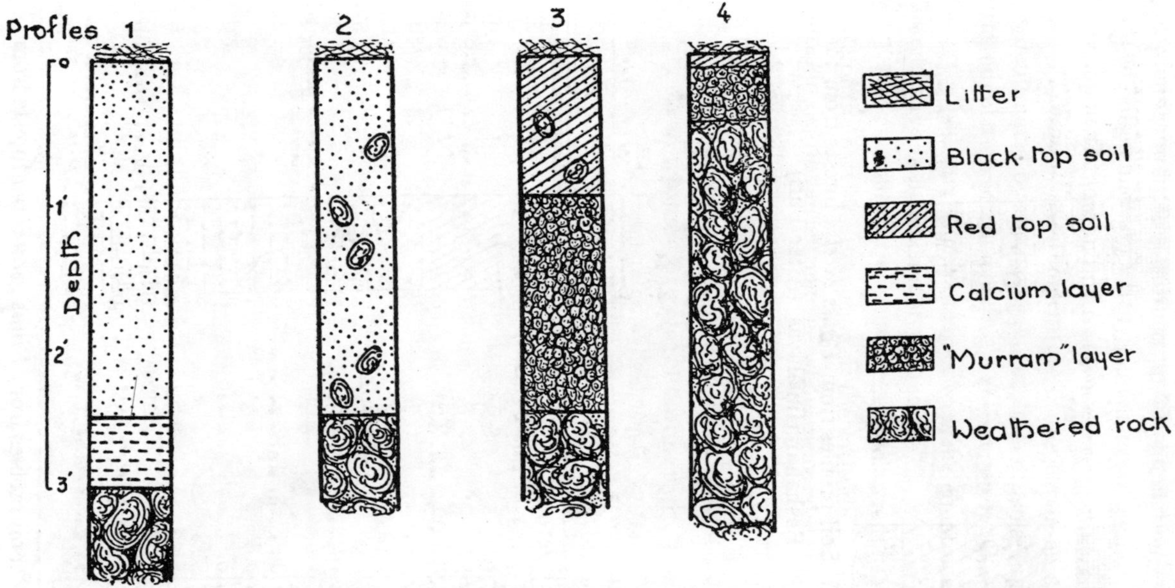

Fig. 30. Profiles showing physical character of soil and parent material of a trap hill bearing open deciduous forest in Madhya Pradesh (After Bhatia, 1955).

1. Black soil in valley with lime concentration at deeper layer.
2. Black soil on midslope without a lime layer.
3. Red loam on plateau.
4. Eroded red soil with 'Murram' layer near surface.

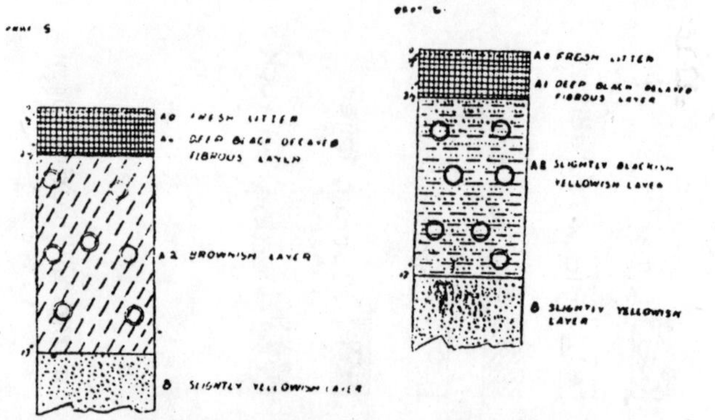

Fig, 31. Soil profiles from *Pinus excelsa* forests from the Bashahar Himalayas (After Puri).

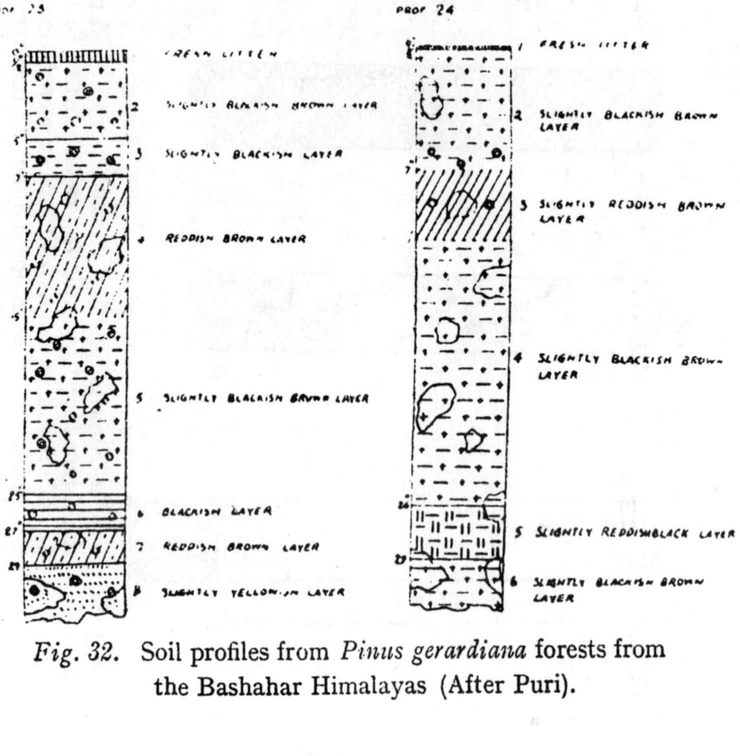

Fig. 32. Soil profiles from *Pinus gerardiana* forests from the Bashahar Himalayas (After Puri).

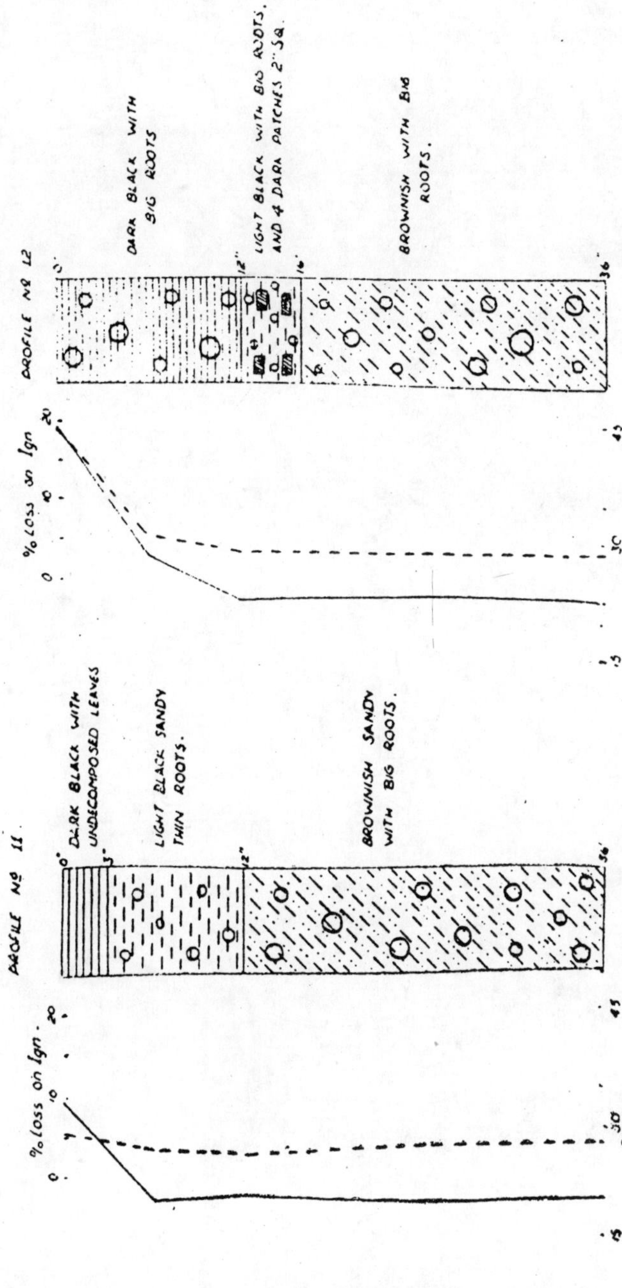

Fig. 33. Soil profiles from sal forests of the Dehra Dun Valley (After Puri).

11

Soil groups of neighbouring regions of India and the World

Some of the large soil groups of India extend to the neighbouring regions of Pakistan, Afghanistan, Burma and Ceylon. In West Pakistan the dry, arid climate has produced thin sandy soils, which are similar in many respects to the soils of Rajputana and south east Punjab. A part of the area in West Pakistan being compound of bare limestones, sandstones and slaty rocks of different geological ages is covered with sandy and siliferous soils; the red, black, lateritic and mountain soils having not been developed in this region.

On account of the more humid climate in East Pakistan and Burma, the soil types formed are similar to those found in wet tropical parts of north-east India. In East Burma and Shan States residual red earthy clay of *terra rossa* type derived from the dissolution of the limestone rock leaving behind its insoluble clay admixture, is present. Lateritic soils of Burma are similar to those of the east coast of the Deccan and Malabar. The forest soils in north Burma and Arakan are not different from wet tropical forest soils developed on the sub-Himalayan Tertiary zone of Assam. Deltaic soils similar in character to those present in the Ganges alluvium occur in central Burma.

In Ceylon the soils are derived from the archean rocks, as in the Deccan, but their correspondence to the geological foundation is less here than in the case of the Deccan soils. In the South-west part of Ceylon the weathering of rocks in wet climate has produced deep profiles and in this respect these soils differ from those of the Deccan. The South Indian red and brown type soils are present in north and east parts of Ceylon.

A. W. Joachim in 1935 had divided the soils of Ceylon into the following main groups :—

(1) Laterite and lateritic soils of the wet S and S. W. of the Island;

(2) Terra rossa soils on the Jaffna limestone;

(3) Steppe soils of the central mountain zone; the wet deep acid soils of the Uva highlands making ideal soils for tea cultivation;

(4) the sandy or gravelly earth, partly of alluvial origin, covering the wide, low plains of the north; and

(5) the marshy (gley) soils of the low country and of the depressions in the mountains.

Black soils, laterites, red soils, podsols and brown earths occur in Australia, Europe, Africa and North America. The Indian black cotton soil, or the regur has been compared to chernozem of Europe. Recent work of Stephen (1949), however, shows that the chernozem soils have more organic matter than the black earths. Therefore, the two are not identical though, they are related.

Podsols and brown earth soils occurring in the Himalayas are related to North European soils, but as pointed out by Hoon (1938) and by other workers, the Himalayan podsols are almost always saturated with bases at the surface and have a less acid reaction than the true podsols of Europe.

Desert soils and soils of the semi-arid regions described from India are also found in Australia, Africa and South America.

A map giving the soil forming rocks of the world is given in Fig. 34.

General characteristics of major soil types of the world, the climatic conditions of the region in which they occur, their drainage, productivity, present use and vegetal cover are given in Table 47 and 47A. Different types of soils and their distribution are shown in Fig. 35.

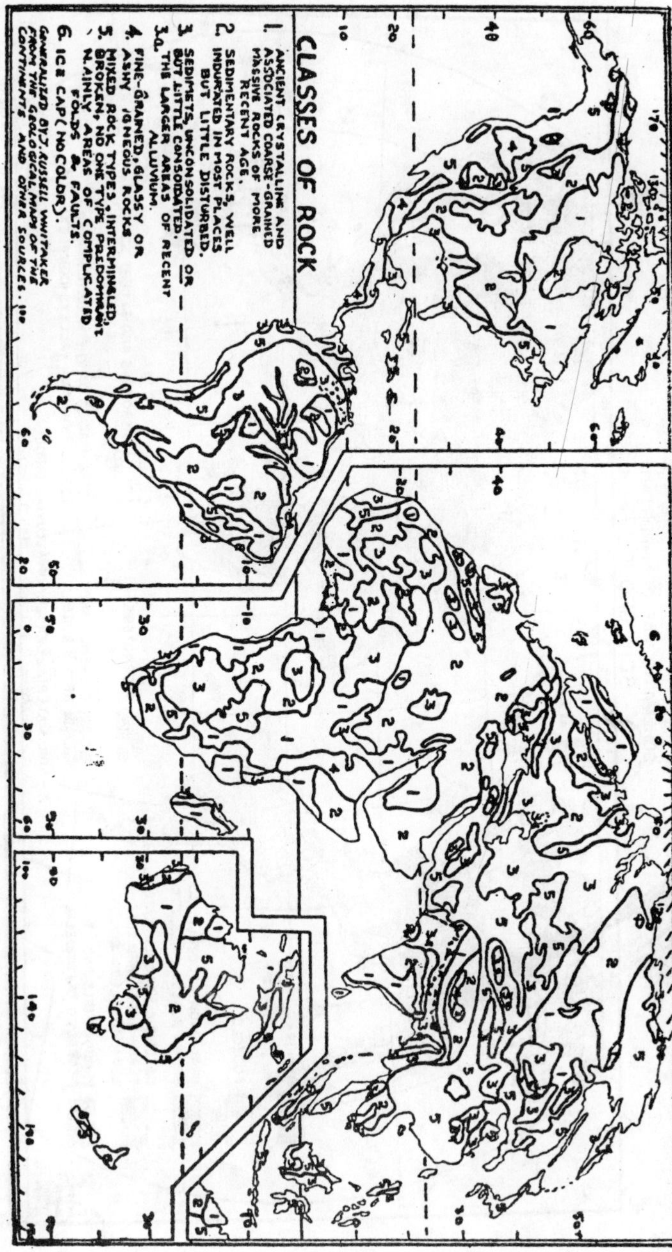

Fig. 34. Map of the world showing classes of rocks.

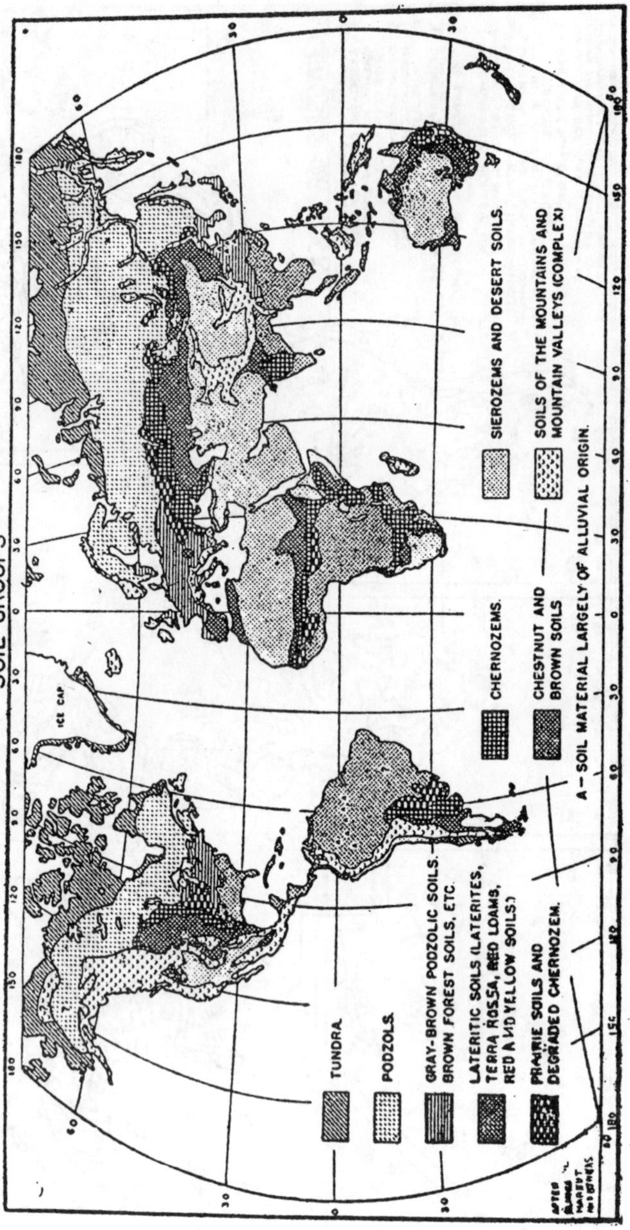

Fig. 35. Soils of the World (From Lutz and Chander 1948)

TABLE 47

GENERAL CHARACTERISTICS OF SOILS AND THEIR ENVIRONS.

(From Soils and Men, U. S. D. A.)

Zonal soils	Profile.	Native vegetation.	Climate.	Natural drainage.	Soil-development processes.	Productivity (crop plants).	Present use.
Tundra	Dark-brown peaty layers over greyish horizons mottled with rust. Substrata of ever-frozen material.	Lichens, moss, flowering plants and shrubs.	Frigrid, humid.	Poor.	Gleization and mechanical mixing.	...	Pasture and a few short-season crops. Hunting and traping are associated enterprises.
Desert	Light-grey or light brownish-grey, low in organic matter, closely underlain by calcareous material.	Scattered shrubby desert plants.	Temperate to cool; arid.	Good to imperfect.	Calcification.	Medium to high if irrigated.	Grazing in large units intensively framed in small units where irrigated. Crops specialized in many places.
Red Desert	Light reddish-brown surface soil, brownish-red or red heavier subsoil closely underlain by calcareous material.	Desert plants, mostly shrubs.	Warm temperate to hot; arid,	do.	do.	do.	do.
Sierozem	Pale greyish soil grading into calcareous material at a depth of 1 ft. or less.	Desert plants scattered short grass, and scattered brush.	Temperate to cool; arid.	do.	do.	do,	do.
Brown	Brown soil grading into whitish calcareous horizon 1 to 3 ft. from surface.	Short grass and bunch grass prairie.	Temperate to cool; arid to semi-arid.	Good.	do.	High if irrigated.	Large farms of small grain (if unirrigated). Ranching in large units.
Reddish Brown	Reddish-brown soil grading into red or dull-red heavier sub-soil and then into whitish calcareous horizon, either cemented or soft	Tall bunch grass and shrub growth.	Temperate to hot; arid to semi-arid.	Good.	do.	Moderate to high, if irrigated. Not suited to dry farming. Grazing good.	Grazing in large units small specialized farms where irrigated.
Chestnut	Dark-brown friable and peaty soil over brown prismatic soil with lime accumulation at depth of $1\frac{1}{2}$ to $4\frac{1}{2}$ feet.	Mixed tall and short grass prairie.	Temperate to cool; semi-arid.	do.	do.	Medium. High where irrigated.	Cereal grains/ especially wheat and grain. Sorghums throughout the world. Excellent grazing in large units.
Reddish chestnut	Dark reddish-brown cast in surface soil. Heavier and reddish-brown or red sandy clay below. Lime accumulation at depth of 2 feet or more.	Mixed grasses and shrubs.	Warm-temperate to hot; semi-arid.	do.	do.	do.	Cereal grains and cotton. Excellent grazing in large units.
Chernozem	Black or very dark greyish brown friable soil to a depth ranging up to 3 or 4 feet grading through lighter colour to whitish lime accumulation.	Tall and mixed grass prairie.	Temperate to cool; sub-humid.	do.	do.	Medium to high. High to very high where irrigated.	Small grains and corn in moderate sized or large units.
Prairie	Very dark brown or greyish-brown soil grading through brown to lighter coloured parent material at a depth of 2 to 5 feet.	Tall grass prairie.	Temperate to cool-temperate, humid.	do.	Calcification with weak podzolization.	High.	Medium to small farm units General farming with emphasis on corn, hogs, and cattle.
Reddish prairie	Dark-brown or reddish-brown soil grading through reddish-brown heavier sub-soil to parent material. Moderately acid.	Tall and mixed grass prairie.	Warm-temperate humid to sub-humid. Possibly some tropical conditions.	do.		Medium to high.	Wheat, oats, corn. cotton, hay and forage crops.
Degraded chernozem	Nearly black A, somewhat bleached greyish A2, incipient heavy B, and vestiges of lime accumulation in deep layers.	Forest encroaching on tall grass prairie.	Temperate and cool; sub-humid to humid.	do.	Caleification followed by podzolization.	Medium t high. Low where strongly degraded.	Agriculture intermediate between chernozem and podzol. Of little importance in the United States.

TABLE 47.—(*Continued*).

Zonal soils.	Profile.	Native vegetation.	Climate	Natural dainage.	Soil-development processes.	Productivity (crop plant).	Preseut use.
Non-calcic Brown (Shantung Brown).	Brown to light-brown friable soil over pale reddish brown or dull-red B horizon.	Mostly deciduous forest of thin stand with brush and grasses.	Temperate or warm temperate; sub-dry, sub-humid to semi-arid.	Good to imperfect.	Weak podzolization and some calcification.	Medium. High where irrigated.	Grazing, dry farming with small grains, specialized irrigated crops including fruits.
Podzol	A few inches of leaf mat and acid humus, a very thin dark grey A_1 horizon, a whitish grey A a few inches thick, a dark or coffee brown B_1 horizon, and a yellowish brown B_2. Strongly acid.	Coniferous or mixed coniferous and deciduous forest.	Cool temperate, except in certain places where the climate is temperate; humid.	Good to imperfect.	Podzolization	Usually low. Medium under good practices.	Small subsistence farms, including dairying, wood lots and pasture important.
Brown Podzolic	Leaf mat and acid humus over thin dark grey A, and thin greyish brown A_2 over brown B horizon which is only slightly heavier than surface soil. Solum seldom more than 24 inches thick,	Deciduous or mixed deciduous and coniferous forest.	Cool-temperate, humid. Effective moisture slightly less than in Podzol.	Good.	do	Low to medium. High where heavily fertilized and well managed.	Small subsistence farms, dairying, wood lots and pasture. Specialised truck crops near large cities are important.
Grey-brown Podzolic	Thin leaf litter over mild humus over dark-coloured surfaces soil 2 to 4 inches thick over greyish-brown leached horizon over brown heavy B horizon. Less acid than Podzols	Mostly deciduous forest with mixture of conifers in places.	Temperate; humid.	do	do	Medium. High where well fertilized and managed.	Small farm units with general farming. Wide variety of crops. Some specialisation. (Much industrial activity).
Yellow Podzolic	Thin dark coloured organic covering over pale yellowish grey leached layer 6 inches to 3 ft. thick over heavy yellow B horizon over yellow-red and grey mottled parent material; acid.	Coniferous mixed coniferous and deciduous forest.	Warm-temperate to tropical; humid.	Imperfect to good.	Podzolization with some laterization.	Poor. Responsive to good management and fertilization.	Small to medium sized farm units. Subsistence crops, cotton, tobbaco, peanuts and some fruit and vegetables. Few live-stock. Many wood-lots and forested areas.
Red Podzolic	Thin organic layer over yellowish brown or greyish brown leached surface soil over deep red B horizon. Parent material frequently reticulately mottled red, yellow and grey; acid.	Deciduous forest with some conifers. (With cogonales-burned over areas covered with cogon, tall coarse garsses).	do	Good.	Podzolization and laterization.	Medium. Responsive to fertilisation and good management.	Small to medium sized farm, with cotton, peanuts, tobbaco, and subsistence crops. Much wasteland and forests.
Yellowish Brown Lateritic	Brown friable clays and clay loams over yellowish brown heavy but friable clays. Acid to neutral.	Evergreen and deciduous broad-leaved trees. Tropical selva. (Some cogonales)	Tropical; wet-dry. High to moderate rainfall.	Good externally. Good or excessive internally.	Laterization and some podzolization.	Low-medium with irrigation and fertilization.	Small farm units, with subsistence and some specialized crops. Some forests.
Reddish Brown Lateritic	Reddish brown or dark reddish-brown friable granular clayey soil over deep-red friable and granular clay. Deep substrata reticulately mottled in places.	Tropical rain forest to edge of savannah (some cogonales).	Tropical wet-dry. Moderately high rainfall.	Good externally and internally.	Laterization with little or no podzolization.	Low to medium. Medium to high where fertilized and irrigated.	Small farm units with subsistence crops. Plantations of citrus, pine-apple, sugarcane etc. Some forests.
Laterite	Red brown surface soil. Red deep B horizon. Red or reticulately mottled parent material very deeply weathered.	Tropical selva and savannah vegetation. (Some cogonales).	Tropical; wet-dry. High to moderate rainfall.	Good externally. Good or excessive internally.	Laterization and a little podzolization.	Low, medium to high with heavy irrigation and fertilization.	Small farm units with wide variety of subsistence crops. Some specialization on plantations. Large areas of wasteland and forest. Mines for iron and aluminium in places.

INTRAZONAL

Intrazonal oils.	Profile.	Native vegetation.	Climate.	Factors responsible for development.	Natural drainage.	Soil development processes.	Productivity (crop plants).	Present use.
Solonchak.	Grey thin salty crust on surface, fine granular mulch just below and greyish friable saltly soil below. Salts may be centrated above or below.	Sparse growth of hallophytic grasses, shrubs and some trees.	Usually subhumid to arid, may be hot or cool.	Poor drainage with evaporation of capillary water. Salty accumulations.	Poor or imperfect.	Salinizition.	Very low except where washed free of salts.	Some grazing. Much waste-land. Used for producing salt and saltpeter in places.
Solonetz.	Very thin to a few inches of friable surface soil underlain by dark hard columnar layer, usually highly alkaline.	Halophytic plants and thin stands of others.	do.	Improved drainage of a sodium solonchak.	Imperfect.	Solonization (desalinization and alkalization).	Low (medium where reclaimed).	Same use as associated normal soils.
Soloth.	Thin greyish-brown horizon of friable soil over whitish leached horizon underlain by dark-brown heavy horizon.	Mixed prairie or shrub.	Usually subhumid to semi-arid. May, be hot or cold.	Improved drainage and leachlng of Zolonetz.	Imperfect to good.	Solodization (dealkalization).	Low to medium.	do.
Wiesenboden (Meadow).	Dark-brown or black soil grading, at a depth of 1 or 2 ft. into greyish and rust mottled soil.	Grasses and sedges.	Cool to warm; humid to subhumid.	Poor drainage.	Poor.	Gleization and some calcification.	Generally high or very high when drained.	Used in connection with associated normal soils. Drained areas similar to Prairie.
Alpine Meadow.	Dark-brown soil grading, at a depth of 1 or 2 ft. into greyish and rust soil streaked and mottled.	Grasses, sedges, and flowering plants.	Cool-temperate to frigld (alpine).	Poor drainage and cold climate.	Pcor.	do.	Yields limited by cool climate.	Mostly summer pasture.
Bog.	Brown, dark-brown or black peaty or much over brown peaty materal.	Swamp forest or sedges or grasses.	Cool to tropical; generally humid.	Poor drainage water covered much of the time.	Very poor.	Gleization.	Low or medium. High in some places where drained.	Special crops when drained. Undrained areas in forest swamp or marsh plants.
Half Bog.	Dark-brown or black peaty material over greyish and rust-mottled mineral soil.	do.	do.	do.	do.	do.	Medium to high when drained. Some low.	Used in connection with normal soils for pasture or forest, where undrained, and for special crops where drained.
Planosols.	Strongly leached surface soils over compact or cemented claypan or hardpan. Some have normal A and B horizons above the claypan or hardpan a secondary profile.	Grass or forest.	Cool to tropical; humid to subhumid.	Flat relief, imperfect drainage and great age.	Imperfect or poor.	Podzolization Gleization. Also laterization in tropics.	Medium to low.	Crops, pasture and forest trees, varying with regions in which they occur.
Ground-water podzols.	Organic mat over very thin acid humus, over whitish grey leached layer upto 2 or 3 ft. thick over brown or very dark brown cemented hardpan or ortstein. Greyish deep substrata.	Forest of various types.	Cool to tropical; humid.	Imperfect drainage and usually sandy material.	do.	Podzolization.	Low to medium in cool areas. Low in warm.	Forest. Some land planted to regional crops and pastures.
Ground-water laterites.	Grey or grey-brown surface layer over leached yellowish grey A2 over thick reticulate mottled cemented hardpan at a depth of 1 ft. or more. Hardpan upto several fcet thick. Laterite parent material. Concretions throughout.	Tropical forest.	Hot and humid; wet and dry seasons.	Poor drainage and considerable or great age.	do.	Podzolization and laterization.	Low to medium.	Subsistence crops, sugarcane and forest trees.
Broad forest (Braundrde).	Very dark brown friable surface grading through lighter colorcd soil to parent material. Little illuviation. High absorbed calcium.	Forest, usually broad-leaved,	Cool-temperate to warm temperate humid.	High calcium coloids and youth.	Good.	Calcification with very little podzolization.	High.	Subsistence and regional crops, pasture and forest.
Rendzina.	Dark greyish brown to black granular soil underlain by grey or yellowish usually soft, calcareous material.	Usually grassy. Some broad leaved forest.	Cool to hot humid to semi-arid.	High content of available lime carbonate in parent material.	do.	Calcification	do.	Regional and special crops. Pasture grasses.

AZONAL

Azonal areas.	Profile.	Vegetation.	Climate.	Drainage.	Soil development processes.	Productivity and present use.
Lithosols.	Thin, stony surface soils, little or no illuviation. Stony parent materials.	Depends on climate.	All climates. Most characteristic of deserts least so of humid tropics.	A wide range mostly god to excessive.	Those characteristic of the region. Little effect has been made.	Forestry, grazing barren. Some agriculture on limited areas. Low productivity.
Alluvial soils.	Little profile development. Some organic matter accumulated. Stratified.	do.	All climates except extremely frigid ones.	A wide range, mostly poor to good.	do.	Practically all crops of world represented. Yields vary from very high to very low. Large proportion of world's population supported by production from alluvial soils. Both subsistence farms and large plantations on there soils.
Sands, dry.	Essentially no profile. Loose sands.	Scanty grass of scrubby forest. Much of land has no vegetation.	Humid to arid, temperate to hot.	Excessive.	do,	Very seldom used except for grazing.

At the 4th International Congress of Soil Science at Amsterdam the following classification given in Table 48 of tropical and sub-tropical soils was proposed.

TABLE 48

Classification of tropical and sub-tropical soils.

Order.	Sub-order.	Great soil groups.	
Zonal.	Humid regions.	1. (a) Podzolic yellow soils with black humid top soil.	
		(b) Red yellow podzolic soils.	
		2. Non-laterised red soils $\frac{SiO_2}{R_2O_3} > 2$	
		3. Laterised red soils $\frac{SiO_2}{R_2O_3} < 2$	
		4. Degraded grey and black clay soils.	Catenary association.
		5. Illuvial grey and black clay soils.	
	Arid regions.	6. Red desert soils.	do.
		7. Dark desert soils.	
Inter-zonal.	Calcimorphic	8. Red calcareous soils over lime stone.	do.
		9. Black calcareous soils over lime stone.	
		10. Red calcareous soils over marls.	do.
		11. Black calcareous soils over marls.	
	Halomorphic	12. Saline soils.	
	Hydromorphic	13. Ground water lateritic soils.	
		14. Planosols.	
		15. Organic soils.	
Azonal.		16. Lithosols.	
		17. Regosols (ash, loess, dry sands).	
		18. Alluvial soils.	

The taxonomic arrangement of soils showing the way in which they are the expression of varying degrees of zonality, hydromorphism, halomorphism, calcimorphism and ferrimorphism is given in Fig. 36, after Stephen (1949).

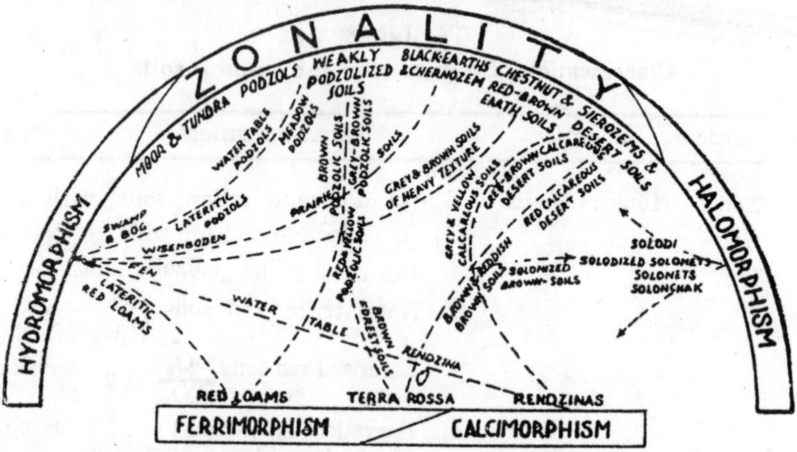

A TAXONOMIC ARRANGEMENT OF SOILS.
(after Stephens, 1949.)

Fig. 36.

Although the parent material has a good deal of influence on the type of soil that will be produced, the effect of climate in determining the character of the profile developed is no less small. This is shown in Table 49.

TABLE 49

Relationship of calcimorphic soil to climatic features and parent material.

Parent material.	Climate			
	Humid.	Sub-humid.	Semi-arid.	Arid.
Soft or powdery calcareous material.	Rendzina.	Rendzina.	Grey calcareous soils.	Brown-grey calcareous desert soils.
Hard or nodular calcareous material.	Brown forest soils.	Terra rossa.	Brown soils, reddish-brown soils, solomized brown soils.	Red calcareous desert soils.

Soil in fact is the product of both climate and parent material, and in the forest and grassland the effect of vegetation is extremely important.

12

Environment of Shoot

The atmosphere is the natural environment of the shoot except for aquatic plants which too require the same gases and conditions for growth through the medium of water. Carbon—the chief element of the plant body is fixed up in light by green tissues from carbon dioxide of the atmosphere. Together with it the metabolic level of the shoot is conditioned by all the factors which constitute and influence the atmosphere. Besides these factors the reactions of the organisms upon each other and biotic exploitation are responsible for large scale influences on the distribution of plants. Hence, climatic and biotic factors both constitute primarily the shoot environment.

The climate and its effect upon vegetation shall be considered first and it will be followed by a study of light, temperature and air conditions upon growth and distribution in separate chapters.

Climate

The chief elements of climate that influence the growth of plants regionally are water and temperature, though, light, frost and wind may locally become limiting factors in the proper development of some species.

The radiant energy of the sun is responsible for the climate of any region. The ultimate source of radiation energy or insolation is the sun, whose diameter is more than 100 times that of the earth. The approximate temperature of the sun is estimated to be of the order of 10,000 degrees F. Radiant energy reaches the earth in the form of waves. The visible part of it is called light and the invisible part produces heat, ultra violet and other rays. All the energy of the sun does not reach, nearly 6-8% is absorbed by the atmosphere and as much as 40% may be reflected by the clouds. The amount of energy ultimately received by the earth is governed by a variety of factors. It can be measured directly by instruments. The sun's energy can also be measured by the effects produced on the surface of the earth, such as temperature, precipitation, winds and air pressure. These resultant manifestations of the sun's energy differ from place to place and also vary greatly with time, producing diurnal and seasonal differences in weather and climate. The weather of any place is "the sum total of its atmospheric condition (temperature, pressure, wind, moisture and precipitation) for a short period of time. Climate on the other hand, is a composite or generalisation of the variety of day to day weather conditions". Kendrew (1922) states that "certainly no picture of climate is at all true unless it is painted in all the colours of the constant variation of weather and the changes of season which are the really prominent features". The controls and elements of weather and climate, modified from Finch and Trewartha (loc. cit.) are given in Table 50.

TABLE 50

The controls and elements of weather and climate

Source.	Climatic controls.		Climatic elements.		
Sun's radiant energy.	1. Latitude. 2. Land, or water. 3. Winds and air masses. 4. Altitude. 5. Semi-permanant low and high pressure centres. 6. Ocean currents. 7. Storms.	**Acting upon**	1. Temperature. 2. Precipitation and humidity. 3. Winds and air pressure.	**Produces**	Types and varieties of weather and climate.

The amount of energy received by the earth and its resultant effects are, although controlled by the 7 factors given above latitude is the most important single factor that determines climate of region. On account of the inclination of the earth's axis the maximum energy is received at the equator, decreasing with distance towards the poles. This is so in spite of the fact that the maximum length of the day increases from the equator towards the poles as shown in Table 51.

TABLE 51

Duration of maximum day and night at given latitudes.

Latitude.	Duration of maximum day length (also night) in hours.
0°	12
17°	13
41°	15
49°	16
63°	20
66°	24
67° 21′	one month
69° 51′	2 months
78° 11′	4 months
90°	6 months

Although, the maximum duration of day and night increases from the equator towards the poles the thermal days (A thermal day is the average total daily insolation at the equator) decrease in the same order as shown in Table 52.

TABLE 52

The number of thermal days in a year at given latitudes

Latitude.	Thermal days in the year.
0°	365.2
10°	360.2
20°	345.2
30°	321.0
40°	228.5
50°	249.7
60°	207.8
70°	173.0
80°	156.6
90°	151.6

Due to the differential absorption of sun's energy by bodies of water and earth, different temperatures are produced and these set up air and water currents, which have a great effect on weather and climate of a particular locality. Water surfaces are comparatively less affected by sudden changes in temperature but land surfaces show great variations. This differential heating produces areas of low and high pressures and is the cause of the development of winds, oceanic currents and all the phenomena, including rainfall, that are associated with the climate.

The amount of solar radiation received by the earth varies greatly with the seasons and latitude (Table 53).

TABLE 53

Average amount of the diffuse sky radiation in CAL/CM²/Min. within specified latitudes.

(After Baur and Philips, 1935).

Date	0°-10°	10°-20°	20°-30°	30°-40°	40°-50°	50°-60°	60°-90°N
Dec. 21	0.091	0.079	0.066	0.052	0.034	0.016	0.001
Mar. 21	0.108	0.105	0.099	0.093	0.083	0.066	0.047
June 21	0.105	0.114	0.124	0.125	0.126	0.122	0.153
Sept. 21	0.107	0.104	0.097	0.091	0.881	0.065	0.048

It is not only the incoming radiation that varies, but the outgoing radiation also varies with latitude, as is shown in Table 54.

TABLE 54

The amount of in-coming and out-going radiation at given latitudes.

(After Baur and Philips, 1935).

	0°-10°	10°-20°	20°-30°	30°-40°	40°-50°	50°-60°	60°-90°N
Incoming radiation.	0.354	0.346	0.336	0.297	0.236	0.185	0.145
Outgoing radiation.	0.299	0.299	0.298	0.291	0.269	0.253	0.240
Difference.	0.058	0.047	0.038	0.006	0.033	0.068	0.105

Thus, there is a net gain of solar energy at low latitudes and a net loss at high latitudes. This induces the development of winds, and by the setting up of circulation, temperatures are equalised and abnormally high and low temperatures do not develop at the low and high latitudes, respectively.

Temperature—Temperature of the earth as shown above is determined by the net gain or loss of solar radiation and the circulation of the winds. If there were no wind circulation the temperature differences between the equator and the poles would have been much more extreme.

On account of the differences in solar radiations, temperature of the earth is a function of the latitude. Mean temperatures in degrees F. during the whole year and in the months of January and July are given in Table 55 for the various latitudes.

TABLE 55

Mean temperature Deg. F. of the latitude circles.

(After Haurwitz and Austin, 1944).

Latitude.	Year.	January.	July.	Range.
90° N	—8.9	—42.0	30.0	72.0
80° N	—1.0	—26.0	35.6	61.6
70° N	12.7	—15.3	45.1	60.4
60° N	30.0	3.0	57.4	54.4
50° N	42.5	19.2	64.6	45.4
40° N	57.4	41.0	75.2	34.2
30° N	68.7	58.1	81.1	23.0
20° N	77.5	71.2	82.4	11.2
10° N	80.1	78.4	80.4	2.0
Equator	79.2	79.5	78.1	1.4
10° S	77.5	79.3	75.0	4.3
20° S	73.2	77.7	68.0	9 7
30° S	61.9	71.4	58.5	12.9
40° S	53.4	60.1	48.2	11.9
50° S	42.4	46.6	38.1	8.5
60° S	25.9	35.8	15.6	20.2
70° S	7.5	27.7	—9.4	37.1
80° S	—16.6	12.6	—39.1	51.7
90° S	—27.6	7.7	—54.0	61.7

At the Equator the range of temperature is low and it increases from the equator towards the poles. At the poles the range in temperatures may be as high as yearly temperatures at some low latitudes.

In addition, the latitude temperature is governed by altitude and the amount of water. Temperatures are low at high altitudes. Taking average temperature values for the whole world the temperature decreases at the rate of 1° F for a rise of 300 ft. above sea level. This would create complications in representing latitudinal distribution of temperature on the earth as a whole. With a view to show a clear correlation between temperature and latitude a correlation in temperature figures is made by adding 1° F for every 300 ft. of elevation. For drawing isotherms these corrected temperature figures are used and it may be clearly understood that isotherm figures are different from actual temperatures. The differences between isotherm lines and the actual temperature lines in hilly areas will be large as compared to plains.

Water surfaces are not only heated slowly by the same amount of solar radiation, but the loss of heat is also slow. Temperature ranges along the coast are, therefore, smaller than in continental areas.

In addition to this, water currents of different temperature strike the coast and bring about a change in terrestrial temperatures.

Temperature changes bring about pressure differences and set up winds in circulation. The velocity and direction of the winds are affected by changes in temperature. During the month of January, areas of low pressure develop at, or near the equator and areas of high pressure develop at middle latitudes. As a result of this, winds from middle latitudes blow towards the equator, and if they pass over bodies of water first and then over land surfaces these winds bring rains.

During the month of July low pressure areas shift from the equator to middle latitudes in both hemispheres, and high pressure areas are developed locally between the latitudes 20°N and 40°N and between 20°S and 40°S. As a result of this winds blow from high pressure areas towards the low pressure belt and as they pass over the equatorial oceans they invariably bring rain, especially in South-East Asian Countries. This rain is called summer monsoon and is characteristic of these regions.

Wind:—Due to the creation of low pressure areas during summer between 20°N and 40°N, dust storms, gales, etc. develop which become very pronounced in desert areas, producing sand dunes. These dunes are a common feature in tropical deserts.

In continental areas due to pressure differences, orderly winds blow and their importance in the dispersal of seeds, fruits, etc. of trees and their pollination during different times of the year are well known. The winds also disperse tree litter in a forest and determine the direction of forest fires and such destructive phenomena, associated with it. At high altitudes winds cause much destruction in exposed regions to forest crops by uprooting trees, or shaping different types of crown development.

The speed of wind is accurately measured by an anemometer. For the determination of the velocity of wind approximately the following descriptive scale given by Beaufort may be found useful :—

TABLE 56

The Beaufort scale of wind force with velocity equivalents.

Beaufort number.	General dsecription.	Specification for use on land.	Miles per hour.
0	Calm.	Smoke rises vertically.	Less than 1.
1	Light air.	Wind direction shown by smoke drift but not by vanes. ...	3 to 1.
2.	Slight breeze	Wind felt on face, leaves rustle; ordinary vane moved by wind. ...	4 to 7.
3.	Gentle breeze	Leaves and twigs in constant motion, wind extends light flag. ...	8 to 12.

4	Moderate breeze.	Raises dust and loose paper, small branches are moved. ...	13 to 18.
5	Fresh breeze.	Small trees in leaf begin to sway; crested wavelets form on inland water.	19 to 24.
6	Strong., breeze.	Large branches in motion whistling heard in telegraph wires. ...	25 to 31.
7	Moderate gale.	Whole trees in motion.	32 to 38.
8	Fresh gale.	Twigs broken off trees, progress generally impeded. ...	39 to 46.
9	Strong gale.	Slight structural damage occurs, chimney pots removed. ...	47 to 54.
10	Whole gale.	Trees uprooted, considerable structural damage. ...	55 to 63.
11	Storm.	Very rarely experienced; wide spread damage. ...	64 to 75.
12	Hurricane.	...	above 75.

The direction of winds which is so important from the point of view of the distribution and pattern of rainfall, is different at different latitudes, both in the Northern as well as in the Southern hemispheres. A summary of the wind and pressure belt at the earth's surface is given in Table 57.

TABLE 57

Summary of the wind and pressure belts at the earth's surface.

(After Hann and Conard).

Lat.	75°N	60°N	30°N	10°N	Equator	10°S	30°S	60°S	80°S
Wind.	ENE	WSW	NE	ENE	ESE	SE	WNW	WNW	ESE
			calms		calms		calms		
Press. mb.	1014	1010	1016	1010	1010	1012	1018	990	997

Precipitation :—

The most important effect of winds is that they carry moisture from ocean surfaces and bring rain in different continents; or if they are dry they bring about evaporation. The direction of the winds and the consequent rainfall or desiccation, which they may bring on land surface, depend upon a number of factors, such as latitude, altitude, proportion of land and water, etc. The annual precipitation over the earth is more generally a function of the latitude. The zonal distribution of annual precipitation over oceans and continents in the two hemispheres is given in Table 58.

TABLE 58

Zonal distribution of annual precipitation, in inches over the earth.

(After Haurwitz and Austin, 1944).

Latitude.	90°–80°	80°–70°	70°–60°	60°–50°	50°–40°	40°–30°	30°–20°	20°–10°	10°–0°
Northern Hemisphere.									
Oceans. ...	4.4	8.4	26.9	44.2	53,2	46.3	35.3	49.3	78.4
Continents. ...	4.4	5.7	12.0	19.2	20 2	23.2	26.6	32.1	55.3
Weighted mean.	4.4	7.6	16.4	30 2	35.9	34.2	32.0	44.8	73.1
Southern Hemisphere.									
Oceans. ...	1.8	4.0	19,2	42.0	48.0	38.6	36.2	46.6	55.7
Continents. ...	0 7	3.1	6.7	38.1	31.4	22.2	26.0	42.9	60.6
Weighted mean.	0.7	3.3	18 1	42.0	47.6	36.8	33.9	45.8	56.7

13

The role of climate in the development of vegetation and soils:

Since the recognition of water as an important substance and heat as a necessary condition for the successful growth of plants in all phases of their life the role of climatic development in vegetation and soils became fully evident. As early as 1810, Walhenberg, the famous Swedish botanist, distinguished latitudinal zones of plants in western Europe and compared these with the known altitudinal zonation of vegetation in the Alps. These studies helped to show an intimate relationship of plants to temperature changes, and led Alexander von Humboldt to prepare temperature zones of the earth, which was the first map to show the relationship of the temperature factor to the distribution of plants.

Inspired by these discoveries, botanists in western and central Europe started a good deal of experimentation on the effect of temperatures on germination, flowering, etc. of plants. Attempt was made to express this relationship mathematically.

While the effect of temperature in determining the plant growth in Europe, where moisture was available in sufficient quantities during all parts of the year, was correctly stressed, the application of this principle outside Europe, where temperature was uniformly high but moisture conditions during all parts of the year were not sufficient the same was not possible.

In 1806, Dove explained the importance of moisture to

plant development. He stated that "it is evident that if under a given latitude the temperature of the atmosphere is the principle factor, while under another latitude the moisture of the atmosphere is the principle factor, then neither of these should be entirely overlooked but the part played by each must be examined."

In 1867 to 1869, Carl Linsser published very important work containing the effect of temperature on plant growth in one publication and the relationship of moisture with plant development in the other. He showed that taking the world as a whole, the moisture factor was far more important factor. He was the first to establish a ratio of precipitation to temperature and relate this factor with vegetation of the world dividing it into five zones, namely, A, B, C, D and E.

In 1855, A. De Candolle emphasised that temperature was not the only main factor in the development of plants, but the effect of hereditary factor should also be considered.

In 1874, De Candolle gave a physiological classification of plants utilising the information derived from the study of fossil plants. The basis of this classification was that plants are adapted pysiologically to ranges of mean annual temperatures and to drought. His classification has already been mentioned in connection with Raunkiaer's life forms work.

De Candolle's classification was mainly intended for the use of students of past climates and fossil vegetation, therefore, he did not prepare any vegetation map, which work was, however, left to Oscar Drude who divided the Northern Hemisphere into the four following zones:—

(1) North glacial zone;

(2) Northern winter cold zone;

(3) Northern summer hot zone;

(4) Tropical zone.

This map does not give much relationship of climates with the vegetation, but several years earlier to Drude, Grisebach in 1866 had already prepared the first map of the world vegetation on the climatic side. Dove had published in 1848 a mean monthly temperature map of the world; and in 1879

Alexander Supan had prepared a map showing temperature zones of the world.

With this background Köppen, whose name in climatology has become more or less an authority, published in 1884 a small article and a map of earth showing the following temperature zones :—

(1) Tropical (Torrid);
(2) Sub tropical;
(3) Summer hot; temperate;
(4) Winter cold temperate;
(5) Cold;
(6) Polar (Frigid).

In 1890, adopting Köppen's classification, Drude prepared vegetation map of the world basing his types on De Candolle's. Drude's classification is physiognomic and he recognised the following seven main types :—

(1) Forest;
(2) Bush and shrub;
(3) Grass and scrub;
(4) Steppe;
(5) Desert;
(6) Moss and lichens;
(7) Alpine and polar.

The first five types represent a gradation from humid to arid conditions, and the last two to aridity.

In 1898, Schimper, one of the earliest and the greatest plant geographers, studied physiognomy and floristic composition of vegetation and divided world vegetation into three great formations, namely, woodland, grassland and desert. These were further sub-divided into 13 sub-types of vegetation and presented in a world map, indicating density and abundance of plant cover. Later on Schimper and Von Faber (1935) presented 15 plant formation–types that are given below :—

1. Tropical rain forest;
2. Subtropical rain forest;
3. Monsoon forest;
4. Temperate rain forest;

5. Summer green deciduous forest;
6. Needle-leaf forest;
7. Evergreen hardwood forest;
8. Savannah woodland;
9. Thorn forest and scrub;
10. Savannah;
11. Steppe and half desert;
12. Heath;
13. Dry desert;
14. Tundra and cold woodland;
15. Cold desert.

Like Drude, Schimper's types showed gradations from humid to arid climates. Luxuriant tropical rain forest was on one end and desert represented the other end of the classification.

Just two years after Schimper, Köppen's famous map of climatic types of the world was published. Köppen was familiar with the works of Drude and Grisebach on plant geography, but unfortunately he had not seen Schimper's work. He had also seen De Candolle's physiological classification of plants and unfortunately Köppen's classification was based more on this system than on Drude's or Schimper's work in time and if he had based his classification on Schimper's map this would have produced better results. In classifying climate Köppen had used a simple ratio of the monthly rainfall to the saturation vapour pressure corresponding to the mean temperature of the same month. On this ratio the boundary between forest and steppe and between steppe and desert was shown to be 4 and 22, respectively. In 1918, he presented this ratio by the following formula :—

$$I=8P/(5T+120)$$

where P is mean annual precipitation and T is mean annual temperature in F°. By this formula the value of 2 separates steppe from forest.

In 1923, the above formula was modified as :—

$$I=P/T+33$$

and in 1921 the following formula was introduced in the wall map by Köppen and Geiger :—

$$I = P/2t + 14$$

The five principal climatic types of Koppen are presented in Table 59 below.

TABLE 59

Principal climatic types of world

(After Köppen)

Climatic group.		Dry period	Degree of dryness and cold
Tropical rainy climates.	A	f (S)	W
Dry climates.	B	...	S W
Warm temperate rainy climates.	C	f S	W
Cold snow-forest climates.	D	f (S)	W
Polar (snow) climates.	E	...	T F

S=driest period in summer.

W=driest period in winter.

f=constantly moist (sufficient rain or snow) in all months.

TABLE 59

Types of climate

I—Low latitudes (The tropics).

A. Tropical rainy climates.	1. Tropical rain forest	Af, constantly wet; Am, monsoon variety.
	2. Tropical savanna.	(Aw, wet and dry).
	3. Low-latitude dry climates.	
B. Dry climates.	(i) Low-latitudes desert	(BWh, arid).
	(ii) Low-latitudes steppe	(BSh, Semi arid).
	II—Middle latitudes (Intermediate zones).	
	4. Middle latitudes dry climates.	
	(i) Middle latitudes desert	(BWK, arid)
	(ii) Middle latitudes steppe	(BSK, Semi-arid).

The description of the symbols a, b, c, s, w, S, W, etc. used in Köppen's types are given below (from Klagges, 1942):—

a—Mean temperature of the warmest month above 22°.

b—Mean temperature of the warmest month under 22°, at least four months above 10°.

c—Only one to four months above 10°, coldest month above 38°.

d—Temperature of the coldest month less than-38°.

e—Constantly moist (sufficient rain or snow in all months).

g—Ganges type of annual temperature trend, with maximum before the turn of the sun and summer rainy season.

h—Hot, annual temperature above 18°.

i—Isothermal, difference between extreme months less than 5°.

k—Cold winter, annual temperature less than 18°, warmest month above 18°.

k'—The same, but warmest month less than 18°.

l—Mild, all months 10 to 22°.

m—Monsoon rains, primeval forest in spite of one dry period.

n—Infrequent fogs, but high humidity accompanied by lack of rainfall and relatively cool (summer below 24°).

p—The same, with summer temperature 24° to 28°.

p'—The same with very high temperature (summer above 28°).

S—Driest period in summer.

W—Driest period in winter.

S'W'—The same, but rainy season shifted into autumn.

S"W"—The same, but rainy season in two parts, with a short dry season intervening.

u—(Reserved) sudden type of temperature variation with coldest month after summer solistice.

v—Cape verde type of temperature variation with warmest season shifted into autumn.

X—Transition type with early summer rains.

X'—The same with infrequent but intense rain at all seasons of the year.

S—Steppe climate.

W—Desert climate.

Köppen's climate types are represented in world map in Fig. 36 and their relationship with world's vegetation is shown in Fig. 37.

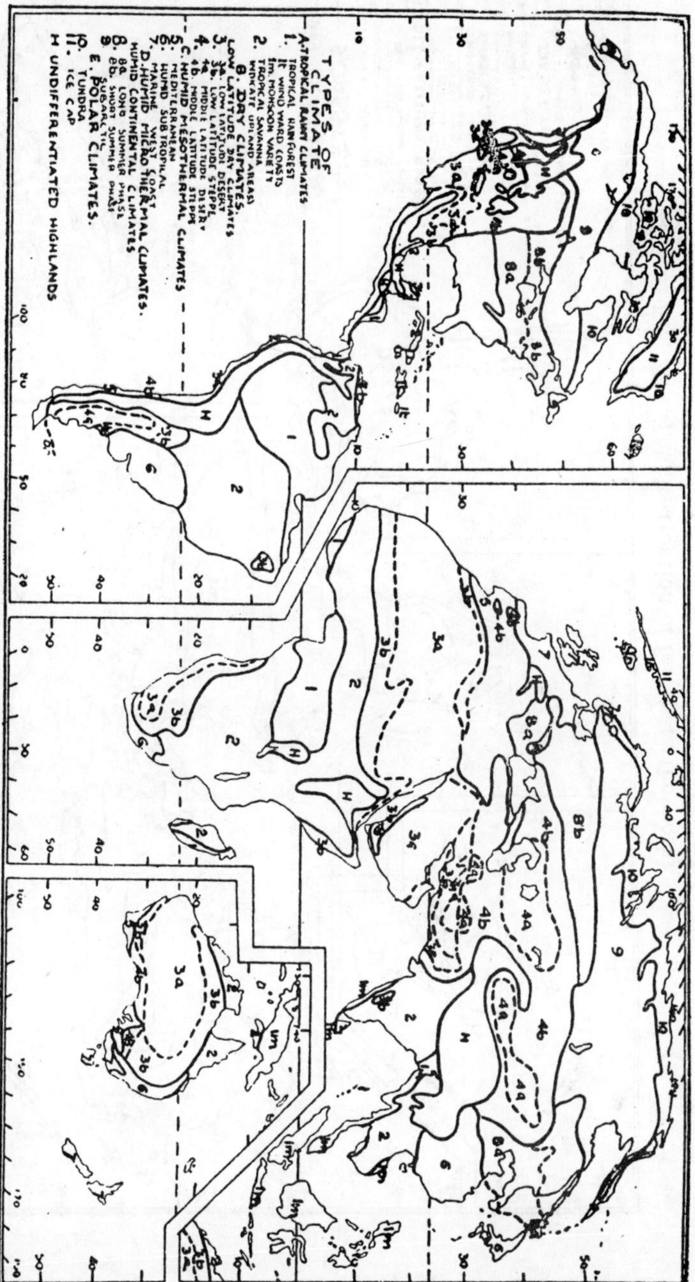

Fig. 36. Köppen's types of climate

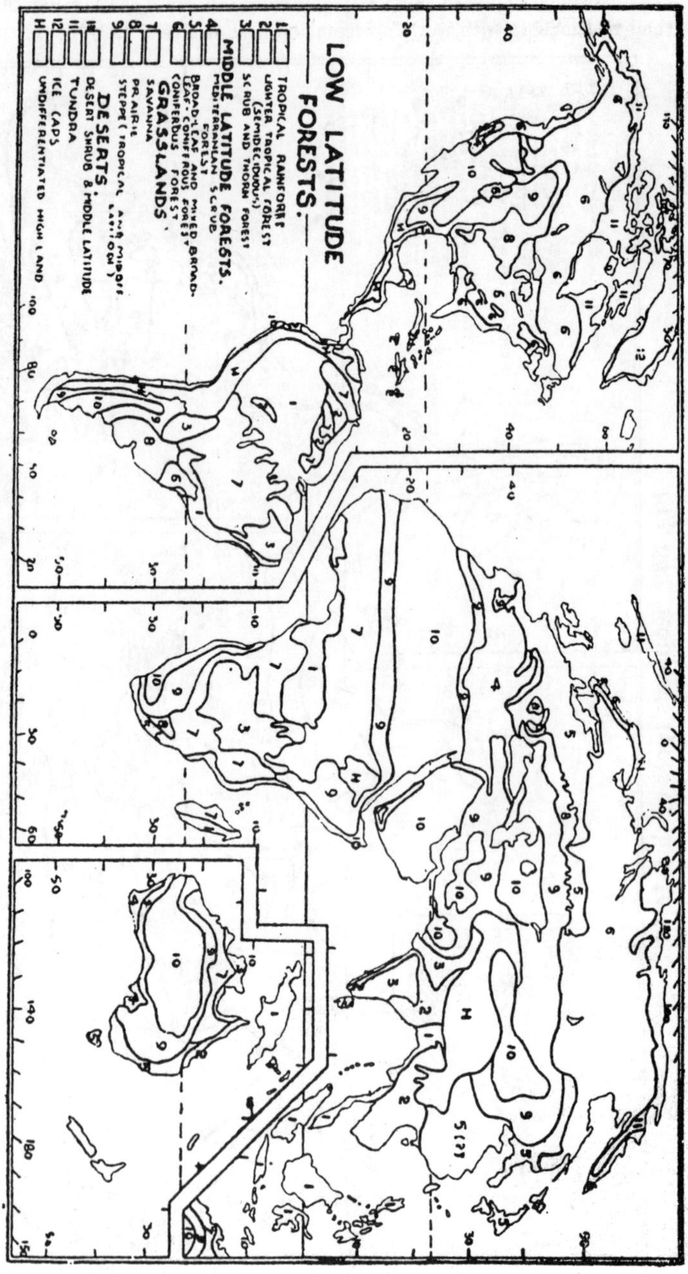

Fig. 37. The distribution of forests in the world.

Precipitation pattern and climograph of his main types are given in Figure 38 and Figure 39 to show the relationship between the various types at a glance.

The relationship between Köppen's climatic types, Schimper and Von Faber's plant formations and Raunkiaer's life forms is shown in Table 60.

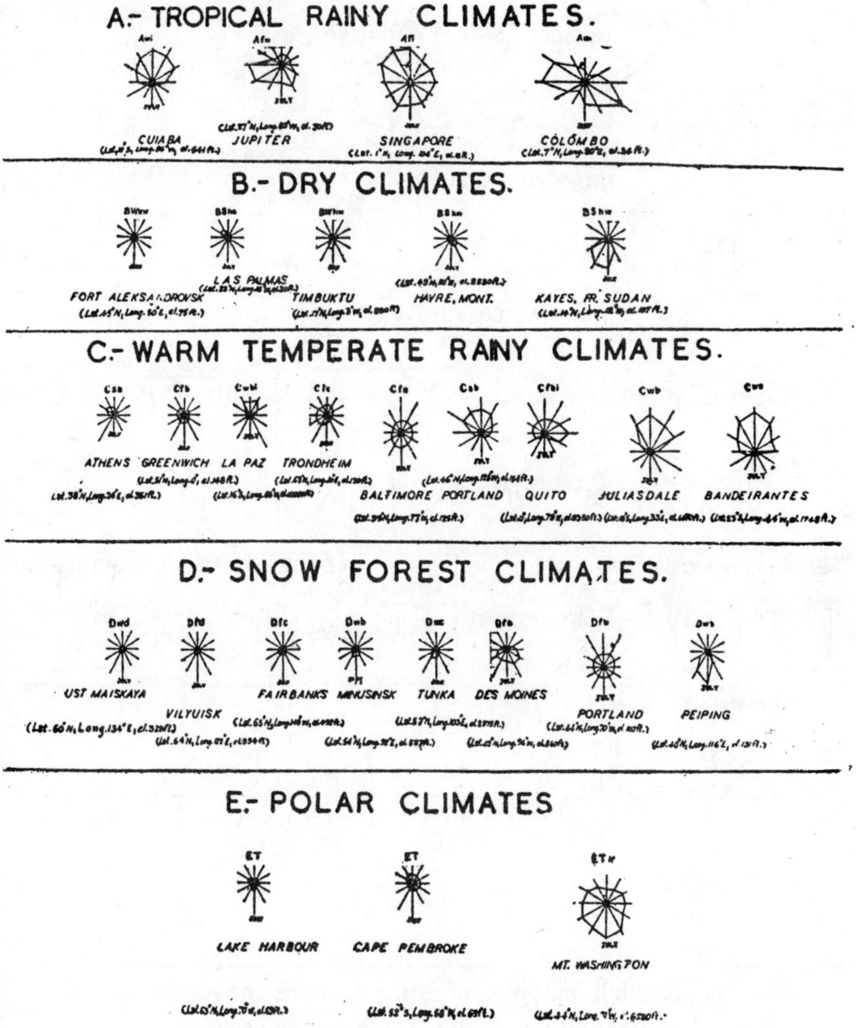

Fig. 38. Rainfall patterns of some typical stations showing the given types of climate.

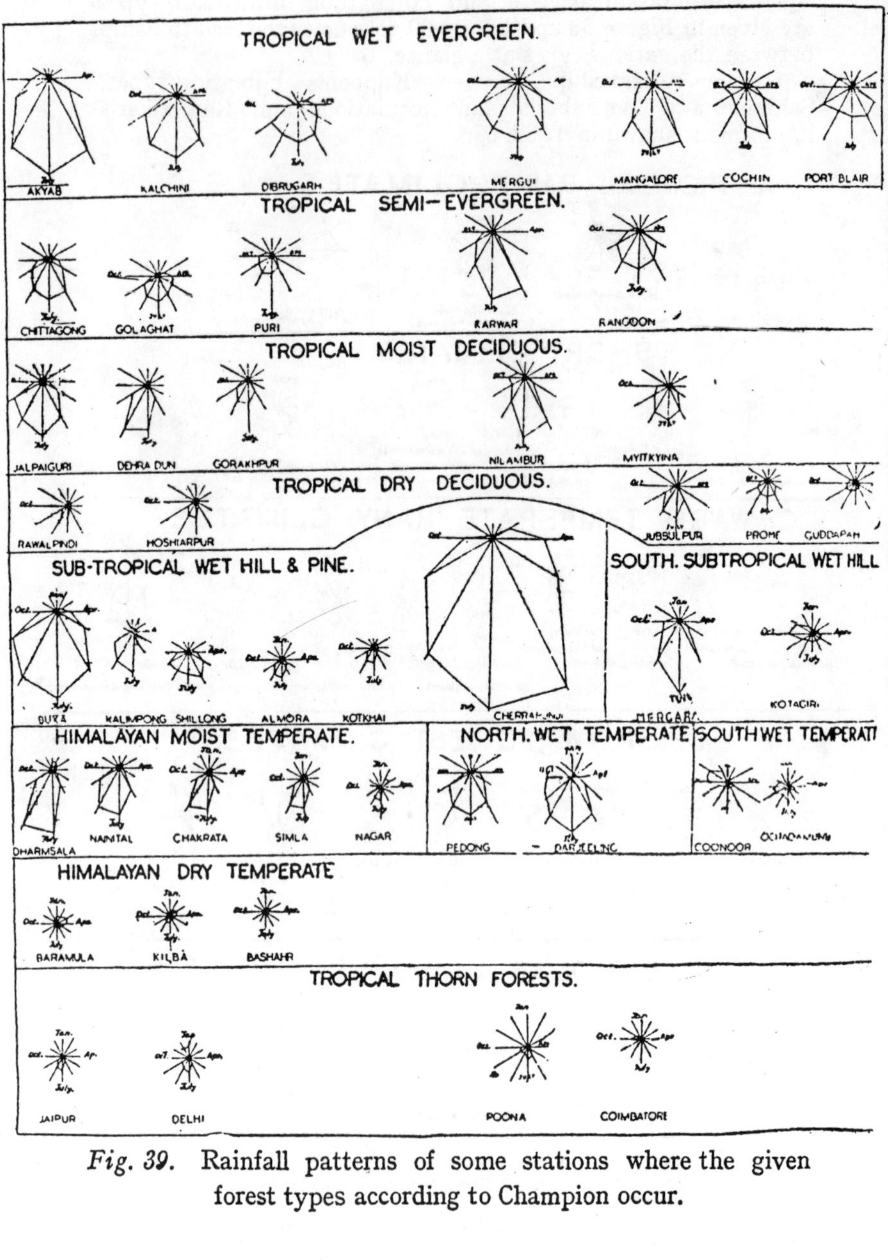

Fig. 39. Rainfall patterns of some stations where the given forest types according to Champion occur.

TABLE 60

Formation-types of Schimper and Von Faber (1935) and their meaning in terms of climate, vegetation and flora. The contacts between one type and one or more others in Schimper and Von Faber's map illustrate the possible transitions, exchanges and genetic derivations. The confinement of a type within one or more of Koppen's climatic areas shows a more or less narrow climatic limitation. The life-forms are Raunkiaer's (1934). (Adapted from Danserau, 1951).

Formation-type of Schimper and Von Faber (1935).	Formation type from which derived.	Contacts with other formation types.	Principal climatic types of Köppen.	Typical area where formation well developed.	Dominant life-form of Raunkiaer (1934) flora.	vegetation.
1. Tropical rain forest.	1	2,3,6,8,10,13,14	Af.	Amazon.	P	P
2. Subtropical rain forest	1	1,3,4,6,7,8,9,10,11, 14	Af, Aw.	New South Wales.	P	P
3. Monsoon forest.	1,2	1,2,6,8,10	Am.	Indo-China.	P	P
4. Temperate rain forest.	1,2	2,5,7,8,10,14	Cw, Cfb.	New Zealand.	H (P)	P
5. Summer green deciduous forest.	4	2,4,6 7,8,9,11,12,14	Cfb. Dfa.	Eastern United States.	H	P
6. Needle leaf forest.	4	1,2,3,5,7,8,9,11,12, 13,14,15	Dfc.	Canada.	H (Ch)	P
7. Evergreen hardwood forest.	2,4	2,4,5,6,8,9,11,13,14	Csb.	Italy.	H (Ch)	P
8. Savannna woodland.	3,4,5,6,7	1,2,3,4,5,6,7,9,10,11	Awi.	Newada.	H (Ch)	P-H
9. Thorn forest and scrub.	7,10	2,6,7,8,10,11,12,13, 14	BShw.	North Eastern Brazil.	Th-P	P-Th
10. Savanna.	7,8	1,2,3,4,8,9,11,14	Awi, Awt.	Senegal, Central Brazil.	H (Ch)	P-Ch
11. Steppe and half desert.	8,10	2,5,6,7,8,9,10,13,14, 15	BSk.	Central United States.	H	H-Ch
12. Heath.	5,6,7	5.6,14	Cfb.	Western Ireland.	Ch-H	
13. Dry desert.	9,10,11	2,6,7,9,10,11,14	BWh.	South Western United States.	Th	Ch
14. Tundra and cold woodland.	6,12	1,2,4,5,6,7,9,10,11, 12,13,15	E.	Ungava.	Ch	Ch-Th
15. Cold desert.	14,11,13	6,11,14	E.	Temperate area highlands and Arctic.	Ch	Ch

The relationship between the plant formations and Raunkiaer's life forms is shown in a bioclimatic diagram (Figure 40) after Dansreau (1951).

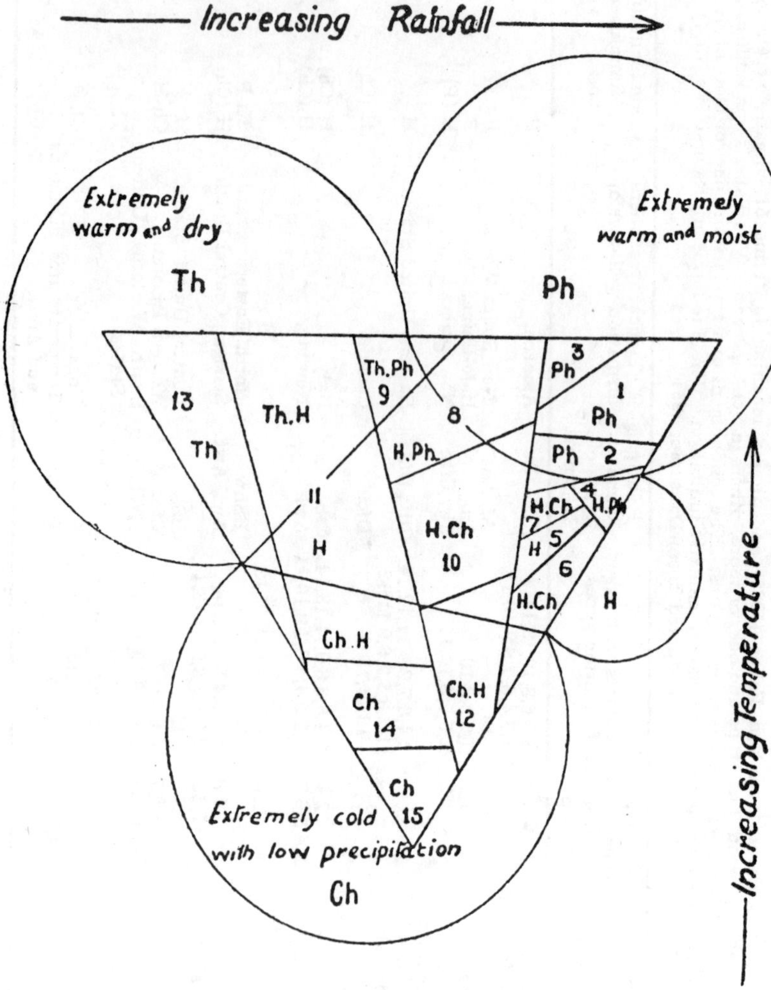

BIO-CLIMATIC DIAGRAM OF DANSREAU, 1951.

Fig. 40

Th. H, Ph., etc. are Raunkiaer's life forms; Nos.1–15 refer to formation types of Schimper and Von Faber given in Table 60. (cn page 223).

Köppen's system of classification of climates was found to be of limited use in tropical and sub-tropical regions. Moreover, this system is not simple, and being unsystematic contains a large number of unnecessary symbols which are hard to put to memory. The ratio given by him is not effective because the vegetation used by Köppen was itself not satisfactory. This classification does not show a clear relationship with the original types of soils. Therefore, from the ecological stand point it is not of very great application, though in number of countries it is still the only system that is used in general classification of climate—vegetation types.

Recently, Kuchler (1949) has presented a physiognomic classification of world vegetation, which is given in Table **61.**

TABLE 61

Kuchler's physiognomic classification of v egetation.

Capital letters.

Woody Vegetation :

B: evergreen broadleaf.

D: deciduous broadleaf.

E: evergreen needleleaf (coniferous).

N: deciduous needleleaf (coniferous).

O: without leaves.

Herbaceous Vegetation:

G: graminoids.

H: forbs.

L: lichens and mosses.

Small Letters.

Group I: Height:

t: tall; minimum height of trees: 25 m. minimum height of herbaceous plants: 2 m.

m: medium tall; trees: 10-25 m., herbaceous plants: $\frac{1}{2}$- 2 m.

l: low: maximum height of trees: 10 m., maximum height of herbaceous plants: $\frac{1}{2}$ m.

s: shrubs: minimum height: 1 m.

z: dwarf shrubs: maximum height: 1 m

Group II: Density:

c: continuous growth.

i: plants usually do not touch.

p: woody plants scattered singly or in groves
herbaceous plants in disconnected patches.

r: rare, yet conspicuous.

b: barren: vegetation largely or entirely absent.

Group III: Special Features:

e: epiphytes.

j: lianas.

k: succulents.

q: cushion plants.

u: palms.

v: bamboos.

w: aquatic vegetation.

y: tree ferns and tuft plants.

This too is found of limited importance in ecological studies.

The most significant climatic factor for the development of vegetation and soils is a direct relationship between precipitation and evaporation and it has been found to be of much greater use than the ratio developed by Köppen. When precipitation and evaporation ratio is high the amount of moisture entering the soil exceeds evaporation and brings about leaching of mineral salts from the surface layer to the subsoil. Under this influence *Pedalfer* type of soil is developed. When the ratio of precipitation to evaporation is low and much of the soil water is evaporated to the atmosphere the development of the soil is characterized by the development of a base-rich layer near, or at the surface of the soil. This type of development is called *Pedacal.* These two types of soils are regional and a close relationship has been shown to exist between these and vegetational development. Köppen's classification does not give us such a relationship which is greatly needed for ecological studies. Another defect in the Köppen's classification is that he makes his types in arid climates on the basis of drainage pattern and mean annual temperature and precipitation factors. His classification does not apply to desert regions as well. This is especially true when Indian conditions are considered.

However, using Köppen's system, Bannerji (1952) has recently tried to interpret conditions in the Indian desert. Applying Köppen's formula, Bannerji gave boundaries of three climates in Rajasthan as follows:—

(a) Dry hot deserts or BWn=0 to 15.22 inches mean annual rainfall.

(b) Dry hot steppe or BSh=15.22 to 30.44 inches mean annual rainfall.

(c) Tropical savanna or AW= 30.44 inches and above mean annual rainfall.

He equated BSh climate with Champion's (1936) "Northern desert thorn forest" which also extends to places with 10″ rainfall. Thus he suggests that boundaries between BWn and BSh be drawn at 10″ isohyet, and that between BSh and Aw at 20″. He equated Aw with Champion's "Southern tropical dry deciduous type".

Köppen's system only gives a broad classification for India and if finer delimitations are required his formulae will have to be modified.

Köppen's system also fails to produce satisfactory results in North Africa and several workers expressed a need for a better index of climate. A suitable classification of climate has been developed by Thornthwaite for North America and the World.

In Australia similar difficulties were expressed in applying rigidly Köppen's system and Prescott and his co-workers have developed a different single value factor of climate with which a good correlation between the development of soil and vegetation has been obtained.

In all the systems the value of temperature and precipitation as important factor in the development of vegetation has been recognised, but Köppen took for his temperature boundaries mean temperature figures of the coldest, or warmest months.

The inherent difficulty in accepting such a principle

in plant-growth is that we have no idea of the effectiveness of the temperature factor. Temperature produces evaporation from the soil and transpiration and evaporation losses of moisture from plant surfaces and, therefore, it effects the liability of precipitation data. The moisture available for plant growth is not merely dependent upon the amount of rainfall, but it is influenced by the amount that is lost in evaporation. In an arid region, therefore, a low amount of rainfall and a high temperature may reduce effectiveness of the moisture to a very low level.

The factor of effective moisture was utilised by Transeau (1905) in marking boundaries of forest centres in Eastern America. He obtained this index of precipitation effectiveness by using the quotient of total annual precipitation and annual evaporation for twelve months.

Later on a rain factor obtained by dividing annual precipitation in millimeters by means of annual temperature in degrees centrigrade was used by Lang in 1915. A map of the world using Lang's factor was prepared. A slight modification of Lang was later on given by De Martonne as follows:—

$$I = P/T - 10$$

In 1926, Alfred Meyer a renowned Climatologist developed N/S quotient by dividing precipitation by saturation deficit. This quotient was extensively used but as the figures of saturation deficit are available only in completely fitted meteorological stations and the records of evaporation measurements are so rare and incomplete, it is difficult to prepare a really definite map. However, a map of India showing Meyer's ratio (Fig. 41) was prepared by Hosking.. A reference to this will be found later on.

Thornthwaite's classification.

To overcome this difficulty of fewer records of evaporation Thornthwaite developed a formula in which there was no need of evaporation data. However, he derived at his formula by a detailed evaporation studies in North America of a number of stations. His formula is given below :—

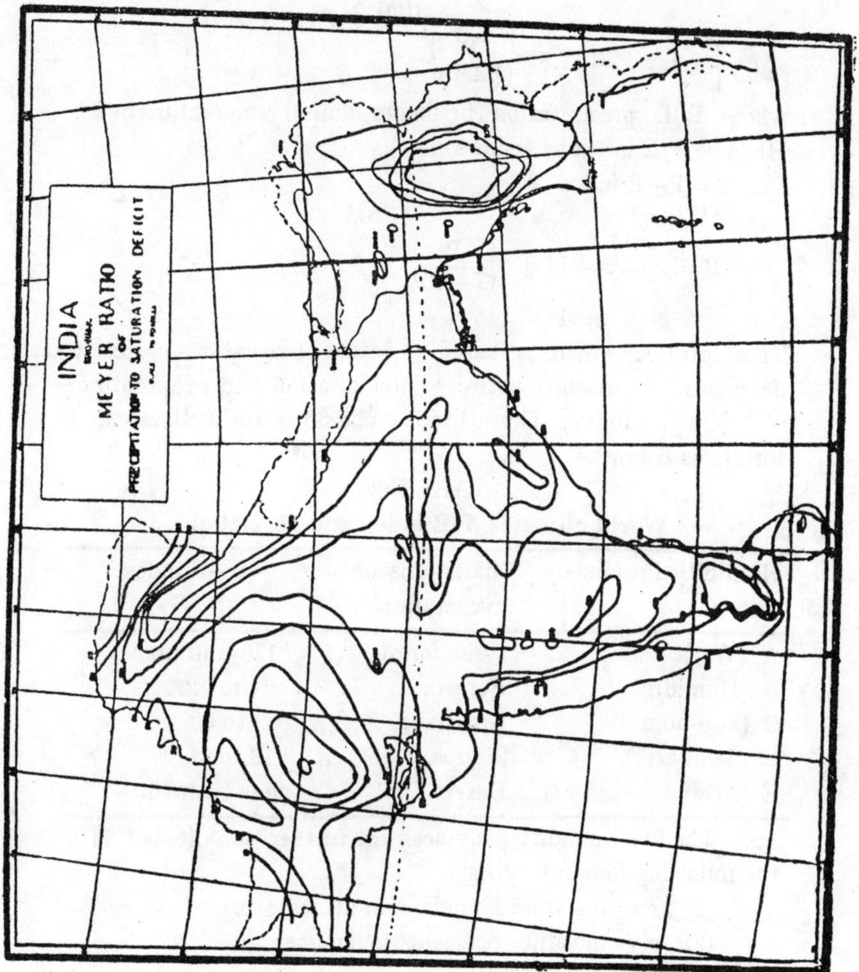

Fig. 41. Meyer's ratio of Precipitation/Saturation deficit.

$$\frac{P}{E} \text{ ratio} = 11.5 \left(\frac{P}{T-10}\right)^{10/9}$$

where P is precipitation in inches, and T temperature in F°. He gave P/E index as follows :—

I=P—E index

$$\text{P—E index} = 11.5 \left(\frac{P}{T-10}\right)^{10/6} n$$

$$n=1$$

This formula which is based on actual observations correctly describes relationship between precipitation and evaporation.

On this index, Thornthwaite classified world climate and forests as below :—

TABLE 62

World climates, P/E index and vegetation.

Humidity Province.	Characteristic vegetation.	P—E index.
A (Wet).	Rain forest.	128 and above.
B (Humid).	Forest.	64 to 127.
C (Sub-humid).	Grassland.	32 to 63.
D (Semi-arid).	Steppe.	16 to 31.
E (Arid).	Desert.	less than 16.

The five humidity provinces are further sub-divided into the following four sub-types.

(1) r = moisture abundant in all seasons.

(2) s = moisture deficient in summer.

(3) w = moisture deficient in winter.

(4) d = moisture deficient in all seasons.

In addition to rainfall, plant growth is governed by temperature changes and Thornthwaite has developed a temperature-evaporation ratio, and temperature-evaporation index as follows:—

$$\text{T—E ratio} = \frac{T - 32}{4}$$

where T is temperature in F° and

$$T - E \text{ index} = 12 \frac{(T-32)}{4}$$

$$n = 1$$

On T — E index Thornthwaite distinguishes six temperature provinces, which correspond to his humidity provinces as follows:—

TABLE 63

Temperature provinces and T — E index.

Temperature Province.	E — T index.
A (Tropical).	128 and above.
B (Mesothermal).	64 to 127.
C (Microthermal.	32 to 63.
D (Taiga).	16 to 31.
E (Tundra).	1 to 15.
F (Frost).	0.

On the basis of summer concentration of temperature efficiency five temperature sub-types have been recognised as follows :—

TABLE 64

Sub-Province.	Percent summer concentration.
a	25 to 34.
b	35 to 49.
c	50 to 69.
d	70 to 99.
e	100.

Thornthwaite's classification is thus an advance on that of Köppen's. He has utilised precipitation effectiveness and thermal efficiency to be of the most important climatic factors, which although were recognised by Köppen, were not clearly represented mathematically by him. The symbols of climate in both these classifications have remained the same, Thornthwaite's map of the world climates is shown in Figure 42.

A relationship between Thornthwaite precipitation effectiveness index and thermal efficiency index with the vegetation types and great soil groups of the world are represented in

Figure 43. A very close correlation between the climate as a chief determining agent in the development of soils and vegetation is clearly seen. It may be mentioned that in this diagramatic representation the effect of other important factors, namely, the parent material, in soil development is not minimised, but if a further sub-division of the soil types could be presented the effect of this factor will be clearly shown.

Thornthwaite's classification is definitely an advance on that of Köppen's and other systems so far as North America and probably also Europe are concernd. It is yet not very certain whether it will apply to delimit Indian forest types proposed by Champion. This has been attempted in the next chapter.

Although, Thornthwaite's P/E index for America is based on actual evaporation measurements these were not taken for other parts of the world. Its chief test will therefore be on India.

Prescott's climatic index for Australia.

For fixing the boundaries of great soil groups and vegetation types in Australia, Prescott (1949) tried the methods given by Köppen, Transeau, Meyer, Lang, Thronthwaite and others but failed to get satisfactory results. Applying the Meyer's ratio of P/s.d for the five major soil groups of Australia, namely, podzols red brown earths, black earths, grey and brown soils and desert formations it was found that both black soils and red brown earths were in the same climatic grouping. Although, seasonal variations in rainfall and drought conditions were seen to be the factors demarcating the boundaries of these types this was not covered in the climatic index.

Prescott therefore, developed the following formula of the climatic index $K = \frac{P}{(s.d.)}\ 0.7.$

where P is the rainfall in inches and s.d. is saturation deficit. On this formula Prescott gave the following boundaries of climatic index of the various soil formations.

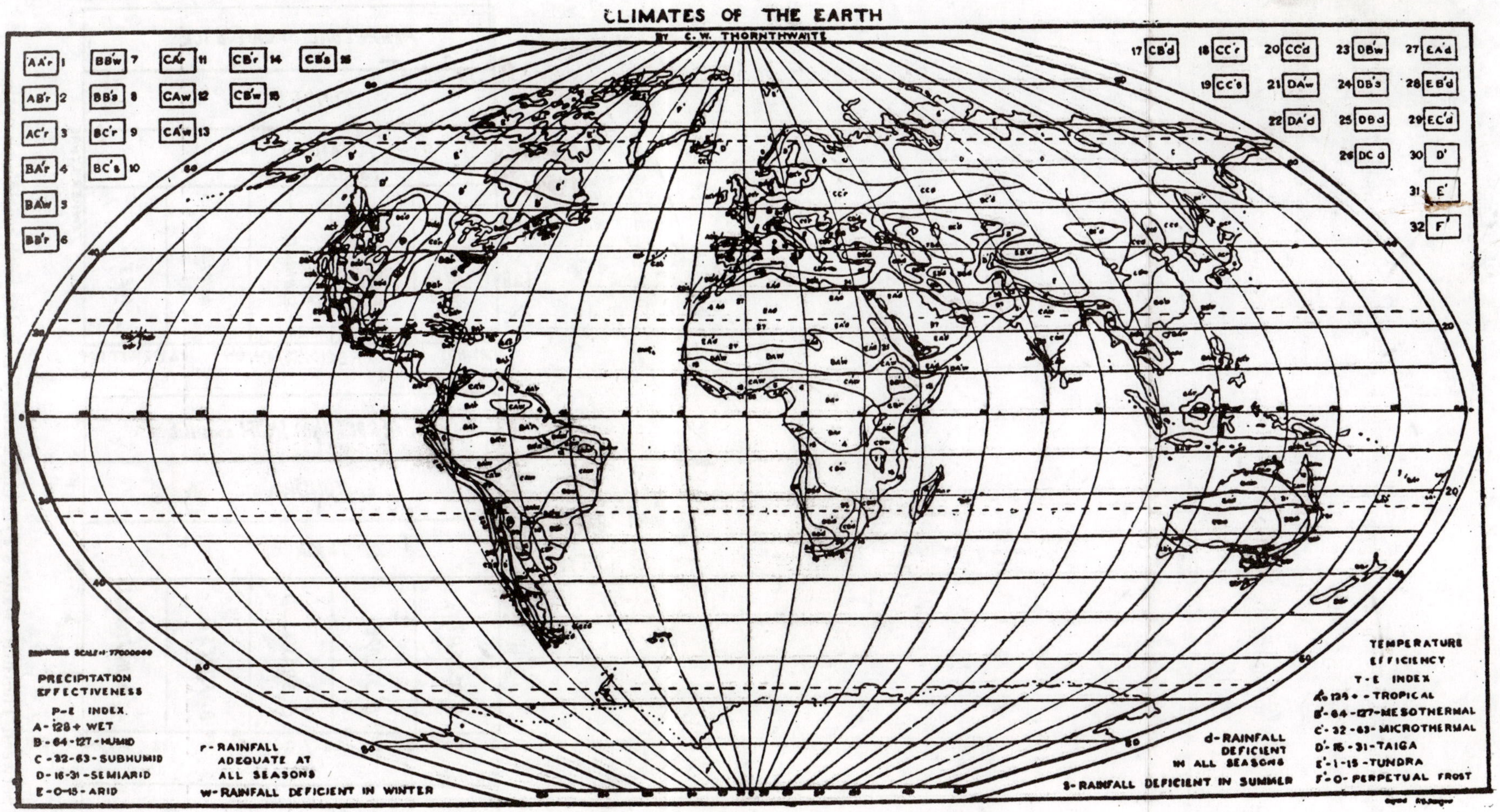

Fig. 42—Thornthwaite's map of world climate.

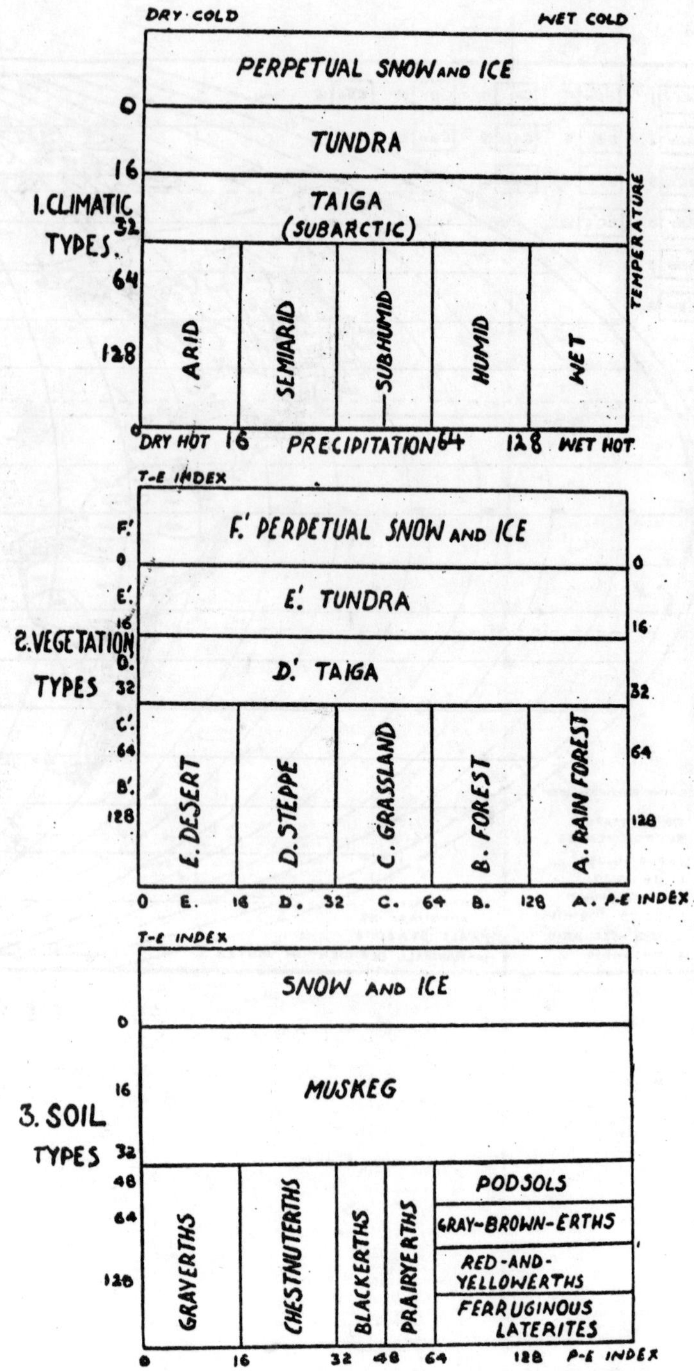

Fig. 43.—Relationship between Thornth waite's P-E index and T-E index with world's climatie typs, vegetation types and soil types.

TABLE 65

Boundaries of climatic index of various soil formations in Australia.

Boundaries.	Climatic index.	Values of index.
Between desert formations and grey brown soils.	23	0.47
Between grey and brown soils and red brown earths and black earths.	45	0.92
Between red brown earths and podzols.	83	1.70

As the measurement of evaporation from vegetation regions is not easy Prescott determined this from free water surfaces and his single value climatic index for the leaching factor in soil formation in Australia is given as P/E^m the value of the constant *m* varies from 0.67 to 0.80 and the mean value is 0.73.

The values of m and the climatic index P/E^m for various types are given by Prescott as in the Table below :—

TABLE 66

Values of m and the climatic index P/E^m

(After Prescott, 1949).

	m	P/E^m	
Geographical boundaries of soil groups in Australia. ...	0.70	0.5	for desert margins for the separation of pedocals and pedalfers.
Records of drain gauges (annual) and natural periods. ...	0.80	1.1	for nil drainage.
Use of water by field vegetation. ... for all examples	0.68	1.3	when rainfall balances transpiration.
for freely transpiring vegetation. ...	0.79	1.5	when rainfall balances transpiration.
Transpiration measurements:			
Lucerne. ...	0.69		...
Wheat. ...	0.67		...
Salt bush. ...	0.76		...

Recently, work has been conducted in other countries. Hurst (1951) applying Thornthwaite's classification to New Zealand has found a correlation between climate and soil development. This is given in Table 67.

TABLE 67

Climatic sub-divisions of New Zealand according to Thornthwaite's rational classification.

(After Hurst, 1951).

Moisture criteria.

Moisture index Im.	Climatic type.
100	A perhumid.
80-100	B_4 humid.
60- 80	B_3 humid.
40- 60	B_2 humid.
20- 40	B_1 humid.
0- 20	C_2 sub humid.
20- 0	C_1 dry sub humid.
40- 20	D semi arid.
60- 40	E arid.

Temperature criteria.

P—E inches		Thermal efficiency type	
>44.8	...	A′	Megathermal.
39.27-44.8	...	B'_4	Mesothermal.
33.66-39.27	...	B'_3	Mesothermal.
28.05-33.66	...	B'_2	Mesothermal.
22.44-28.05	...	B'_1	Mesothermal.
16.83-22.44	...	C'_2	Microthermal.
11.22-16.83	...	C'_1	Microthermal.
5 61-11.22	...	D′	Tundra.
<5.61	...	E′	Frost.

Seasonal variations.

Moist climates.	Aridity index.	Dry climates.	Humidity index.
r, little or no deficiency.	0-16.7	d, little or no surplus.	0-10
s, moderate summer deficiency.	16.7-13.3	s, moderate winter surplus.	10-20
w, moderate winter deficiency.	16.7-13.3	w, moderate summer surplus.	10-20
S_2, W_2, large summer and winter deficiency.	33.3+	S_2,W_2, large winter and summer surplus.	20+

In Table 68, Hurst has compared Thornthwaite's formula with those of Meyer and Prescott.

TABLE 68

Average climatic index of New Zealand

(After Hurst, 1951).

	Yellow-gray soils.	Transitional.	Yellow-brown soils.
Temperature criteria			
Potential evapo-transpiration. (Thornthwaite's rational classification). ...	26.2	26.1	25.8
T-E index (Thornthwaite's empirical classification). ...	61.85	64.78	65.08
Moisture criteria			
Moisture index (Thornthwaite's rational classification). ...	14.2	51.1	112.7
P-E index (Thornthwaite's empirical classification). ...	55.28	80.76	119.0
Meyer ratio. ...	286	450	650
Prescott's index. ...	1.38	2.01	3.03
Aridity index (Thornthwaite's rational classification). ...	4.7	0	0
Humidity index (Thornthwaite's rational classification). ...	17.3	51.1	112.7
Water surplus (Thornthwaite's rational classification). ...	4.58	13.44	29.35
Crowther's leaching factor. ...	14.4	.7	39.3
Seasonal variation criteria			
Months with runoff less than ½ inches (Thornthwaite's empirical classification). ...	8.2	4.4	1.0
Subhumid months (P-E 5.3) (Thornthewaite's empirical classification). ...	8.1	4.2	0.7
Months of low biological activity (Meyer). ...	3.1	0.2	0
Number of rainydays. ...	121	153	165

In 1949, Veen has tried all the known systems of the classification of climates for Netherlands. After comparisons he found the different systems to possess the following merits :—

1. **Meyer—** very plain, provides limits that are acceptable, but does not provide a division of the whole temp-precipitation area.
2. **Köppen—** plain, divides the whole possible temperature—precipitation area, but provides no detailed relationship to soil and forest types.
3. **Cajander—** plain, provides a detailed relation to soil and the forest types, but does not divide the whole possible E/P area.
4. **Rubner—** has no advantages over Meyer, Köppen and Cajander.
5. **Lang—** rather plain, provides no relationship to soil and vegetation types.
6. **De Martonne—**has no advantages over Lang.
7. **Meyer—** no advantages over Lang.
8. **Thornthwaite—**has many advantages.

So the workers that have tried various systems seem to agree to the usefulness of Thornthwaite's system.

14

The Climate of India

The climate of India was first described in detailed maps by Sir John Elliot. Later on Kendrew (1922), C. W. B. Normand (1937) and Stamp (1946) presented shorter accounts, one of which is included in the Imperial Gazeteer of India (1907). The following brief summary will be based on these publications and later publications of the Meteorological Department of India.

The maps for mean pressure and winds, mean relative humidity, number of rainy days, mean annual rainfall, mean maximum temperature are given in figures 44, 45, 46, 47 and 48 from Elliot's climatological atlas, and the sources of the rest of the maps are mentioned in every case.

From the point of view of plant growth the climate of the greater part of India is described by Köppen as tropical rain forest, or tropical savanna, especially that in the east and south of India. A small part of north-west India is described as desert. The chief elements of climate that govern the growth of plants are temperature and rainfall, although locally other factors as frost, winds, etc., may influence vegetation to a certain extent. As the major part of the country has a tropical climate, the temperature conditions are the least limiting factors for plant growth. But rainfall and the degree of moisture are the chief controlling factors in this region and determine the type of vegetation and the extent of its distribution.

In the forests of India there occur a number of tree species and groups that have a wide range of distribution. Their

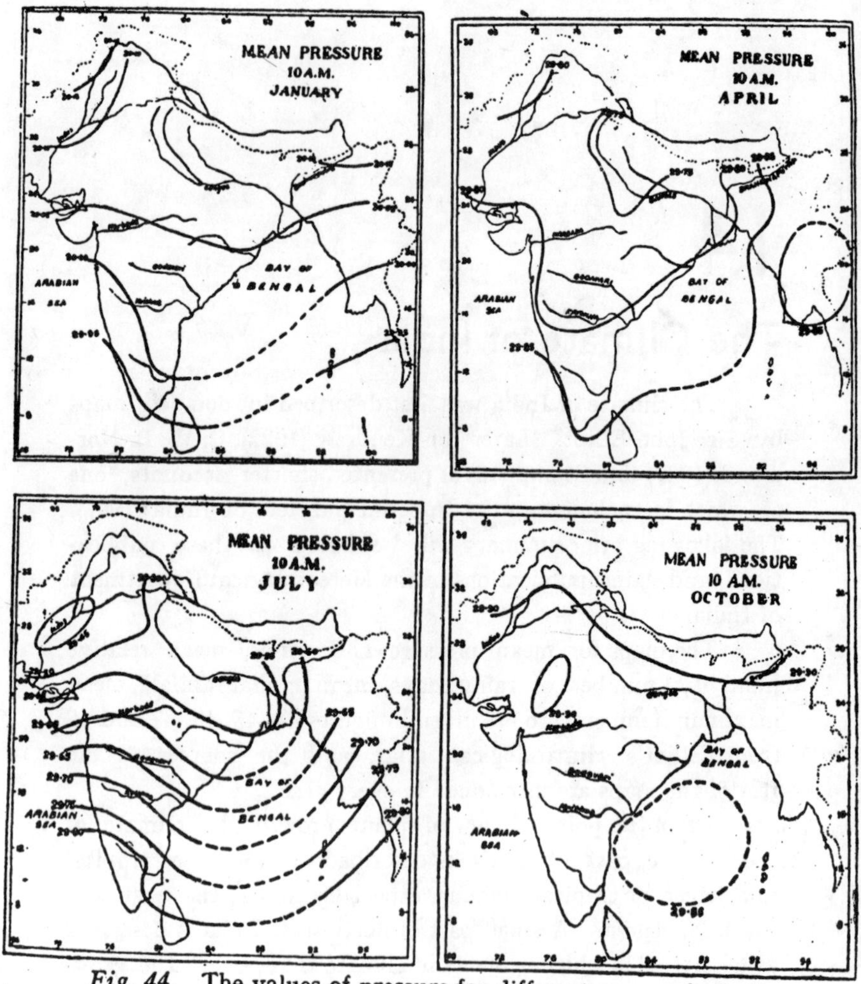

Fig. 44. The values of pressure for different seasons in India.

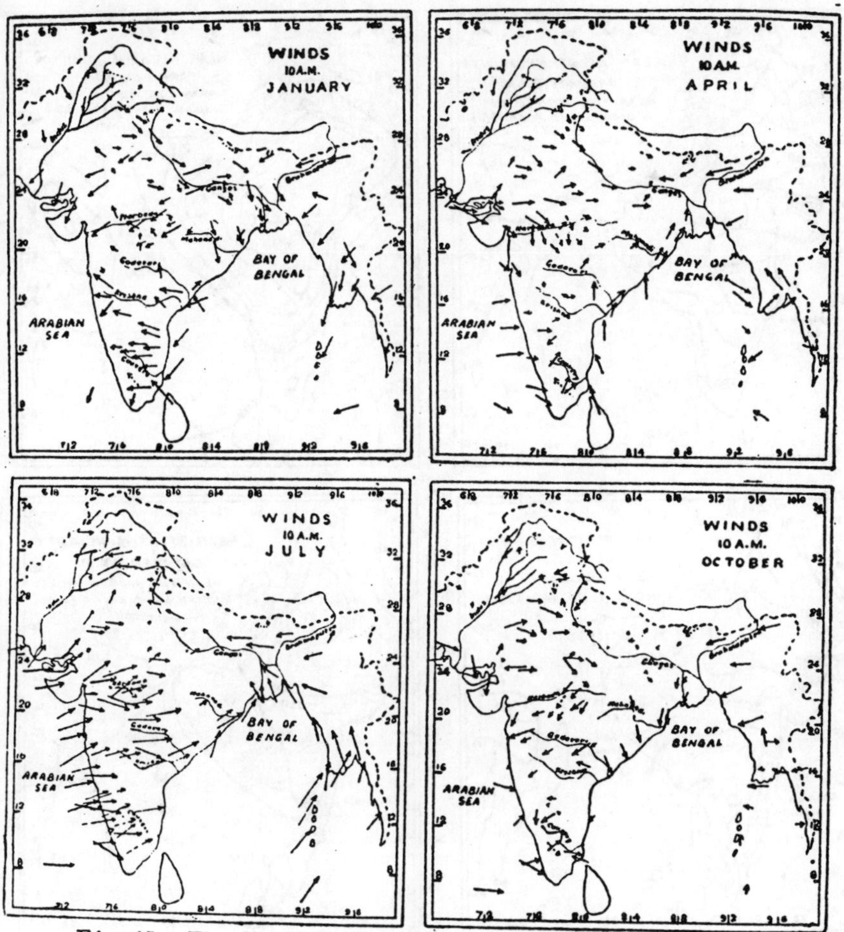

Fig. 45. The direction of winds in different seasons in India.

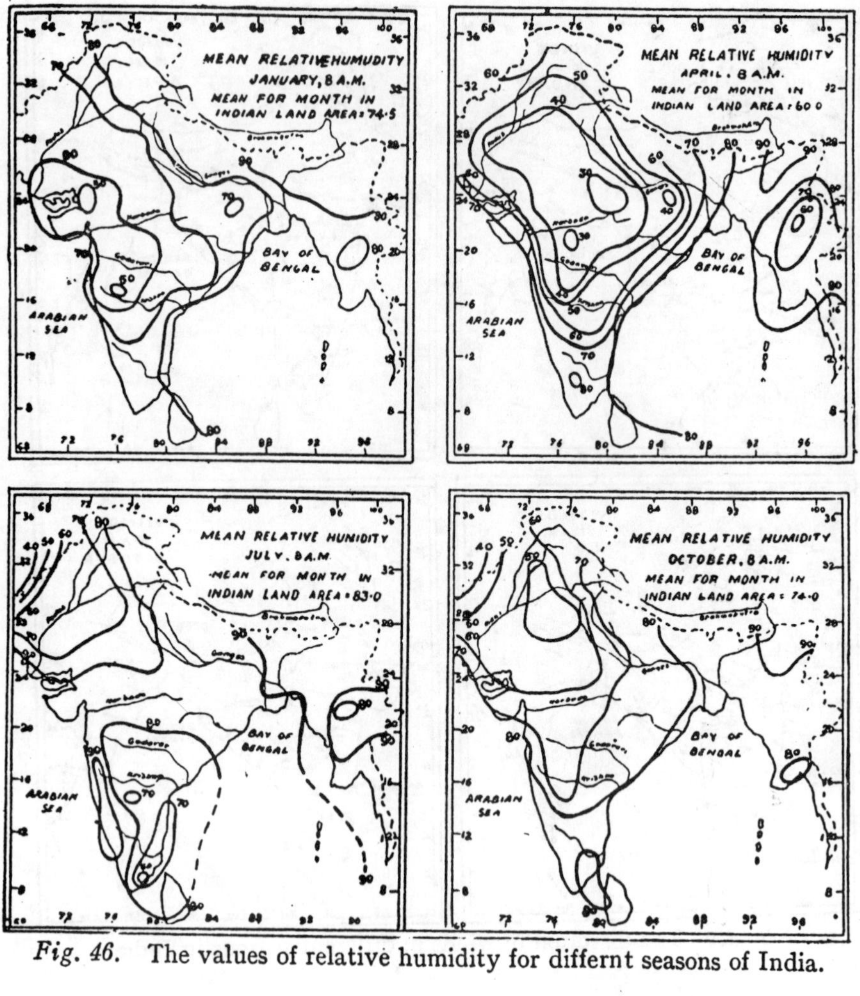

Fig. 46. The values of relative humidity for differnt seasons of India.

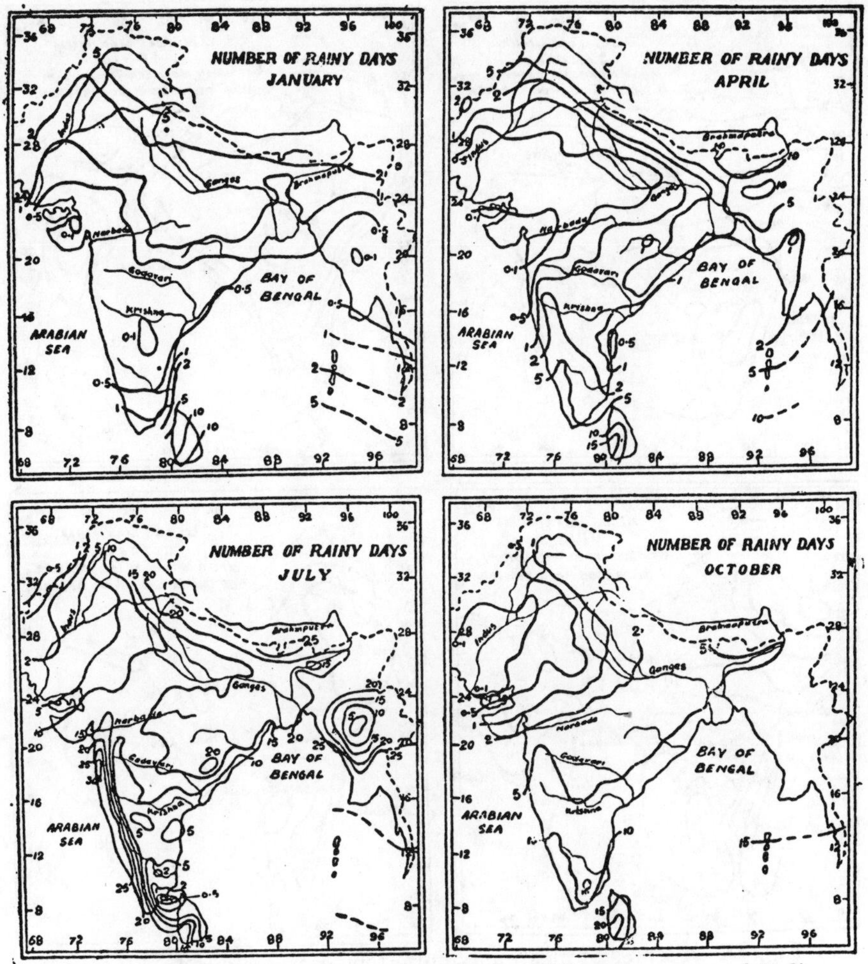

Fig. 47. The number of rainy days during different seasons of India.

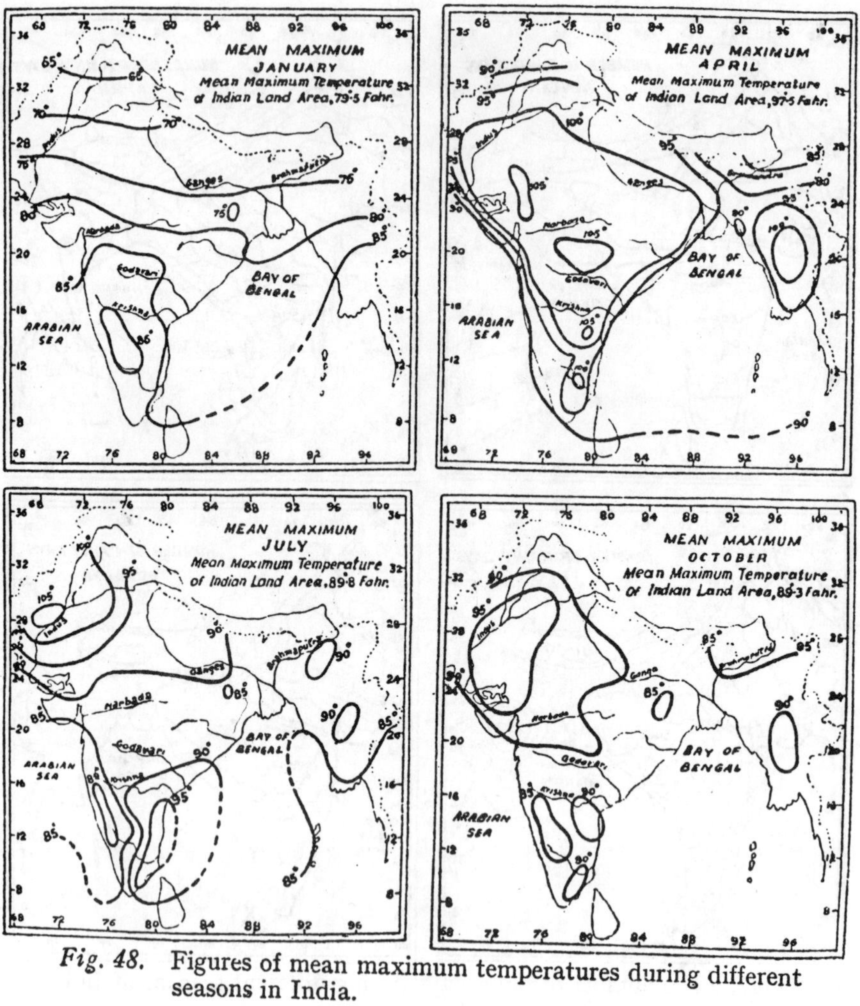

Fig. 48. Figures of mean maximum temperatures during different seasons in India.

number, growth, height, etc. considerably varies in the different zones of rainfall or moisture distribution.

There are also large seasonal variations in climatic conditions and these determine the phenological behaviour of tree species and the periodicity of ground flora vegetation in different tracts. Therefore, seasonal differences in climate are of importance in the study of plants in India. The Meteorological Department of the Government of India have adopted the following four seasons :—

I. The season of the north-east monsoon.

(i) Cold weather season, from January to February.

(ii) Hot weather season, from March to middle June .

II. The season of the south-east monsoon.

(iii) Season of general rain from middle June to middle September.

(iv) Season of retreating monsoon from middle September to December.

(i) Cold weather season.

This season is characterised by the month of January, when mean monthly temperatures rise steadily from the north of India towards the south. For example, the mean January temperature at Peshawar is 50°, in the Ganges valley it is below 60°, at Madras it is 75°, and at Colombo it is 79°. The hills have still lower temperatures.

During this period a feeble high pressure area develops over the colder plains of the north-west India, as a result of which cold winds blow towards the equatorial low pressure area, bringing to large part of the country cold conditions. These winds are influenced in their direction by changes in elevation and blow from the west, or the north-west down the Ganges valley, thence becoming north-easterly over much of the Bay of Bengal. Except in the north-west and eastern coast of Madras, where they bring some rain, these winds are dry. During this period, cyclones originating over the Mediterranean pass through Persia, Baluchistan, or Afghanistan and bring rains to the plains of northern Punjab, which are so very important for wheat and barley crops.

During this season valleys and plains of northern India experience frost damage. Rain is accompanied at lower levels by hail and at higher altitudes in the Himalayas by sleet and ice. Most of the trees in northern India become leafless and ground flora also dwindle.

(ii) **Hot Weather.**

The hot weather over most of India is between March and middle June. April and May over great part of the country are the hottest months. The temperature may be 100^{0}F. or above, except in extreme north-west, where it is still a bit cooler.

In April or May, a feeble low pressure area develops over most parts of India and sea winds blowing from south bring first showers called "Mango showers" to south and south-east India. In drier parts, the winds may cause tornadoes, or whirlwinds. Much damage to vegetation is caused during this period by drought. Tree seedlings and newly planted trees during this period need plenty of watering. Dispersal of tree seeds and litter is brought about during this period by winds. In the desert area, winds are strong and carry dust and sands, forming sand dunes.

(iii) **Rainy Season.**

This starts with the bursting of monsoon winds during the middle of June, or the beginning of July. The low pressure area, which has developed in the hot season in northern India, now extends to the central parts of the country; and it is believed that due to this extension of the low pressure area south-west monsoon winds are caused. The rainfall in Burma is believed to be due to local low pressure area in the heart of the region. Stamp (1946) states that "two things must be emphasised. India is quite cut off from Central Asia, there is no connection between the Punjab low pressure centre and that over the heart of the continent. In the second place, the low pressure area has already been in existence for a month or two before the monsoon arrives. Why the sudden change, which is no mere intensification of the previous light sea breeze Sir John Eliot, late Government Meteorologist of India, believed that during India's hot weather there existed still the normal Equatorial low pressure towards which the south-east trades were drawn and which was separated from the North Indian area by a ridge of high pressure. The equatorial pressure and the ridge of high pressure to the north act as barriers, suddenly they are overcome, and the south-east trade winds are drawn across to the North Indian area." G.C. Simpson is also believed to have been in favour of the above view.

These rain-bearing winds are of nearly twice as much speed as the north-east cold season monsoons and the direction of these is governed by the chief topographical features of the country. The peninsular land mass in the south separates the monsoon into two principal streams. The Arabian Sea stream is less effective and its influence is felt only in the region north of the Gulf of Cambay. This brings some rain to the Punjab plains, but most of the rain falling in the area is delayed being brought by the Bengal stream of wind, which blows as an east wind up the Gangetic plains. The burst of the monsoon therefore, is experienced on different dates in different parts of the country as follows :—

TABLE 69

Mean date of commencement of monsoon

	Mean date of commencement of monsoon.	Mean date of ending of monsoon.
Bombay	June 5	Oct. 15
Bengal	June 15	Oct. 15 to 30
Uttar Pradesh	June 25	Sept. 30
Punjab	July 1	Sept. 14 to 21.

The amount of rainfall and the number of rainy days increase from south-east up the Ganges valley due to the Bay of Bengal stream and from north-west up the Himalayan region due to the Arabian Sea stream. The general rainfall of India month by month from May to October is given in Table 70.

TABLE 70

General rainfall of India from June to October.

		inches
May.	...	3.1
June.	...	7.9
July.	...	11.2
August.	...	10.3
September.	...	7.0
October.	...	3.3
	Total...	42.8

Rainfall for all months in a year in different parts of the country is given in Figure 49.

The greater amount of rain falls during July and August and some rainfall occurs from retreating monsoon winds in September and October. The July rainfall map of India is not very different in pattern from the mean yearly rainfall map of the country. The Meteorological Department of India recognises four rainfall divisions of India, which are not different from the climatic divisions of the country, referred to in the next few pages.

During rainy season rainfall is not regular, but rainy periods are interspersed by periods of bright sunshine and no rain; though, in some parts of the country clouds may hang in the sky for days. There is no regular diurnal variation of rainfall, either. Rainfall may occur from 6 inches to 8 inches or in extreme cases, 15 to 20 inches in an hour at some places. This not only causes serious erosion at the place of occurrence but serious floods in the area. Floods are of common occurrence even in some areas of low rainfall during this period. The occurrence of floods and droughts in N.E., N. W. and Central Peninsular India for the period 1875-1949 are given in Fig. 50.

The vegetation conditions in the rainy season are at their best. Trees show their growth and on account of high temperatures in this region the growth is of a high degree.

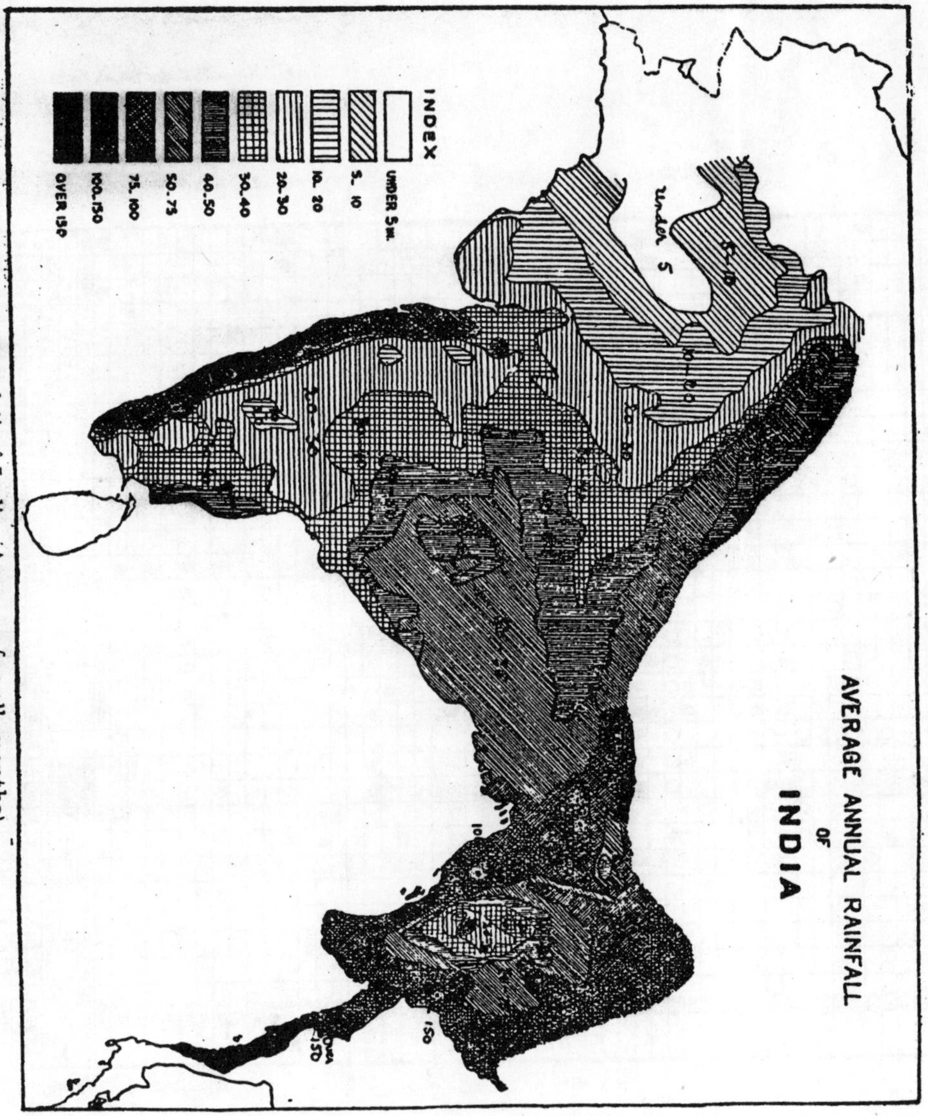

Fig. 49.—Rainfall of India (Average for all months).

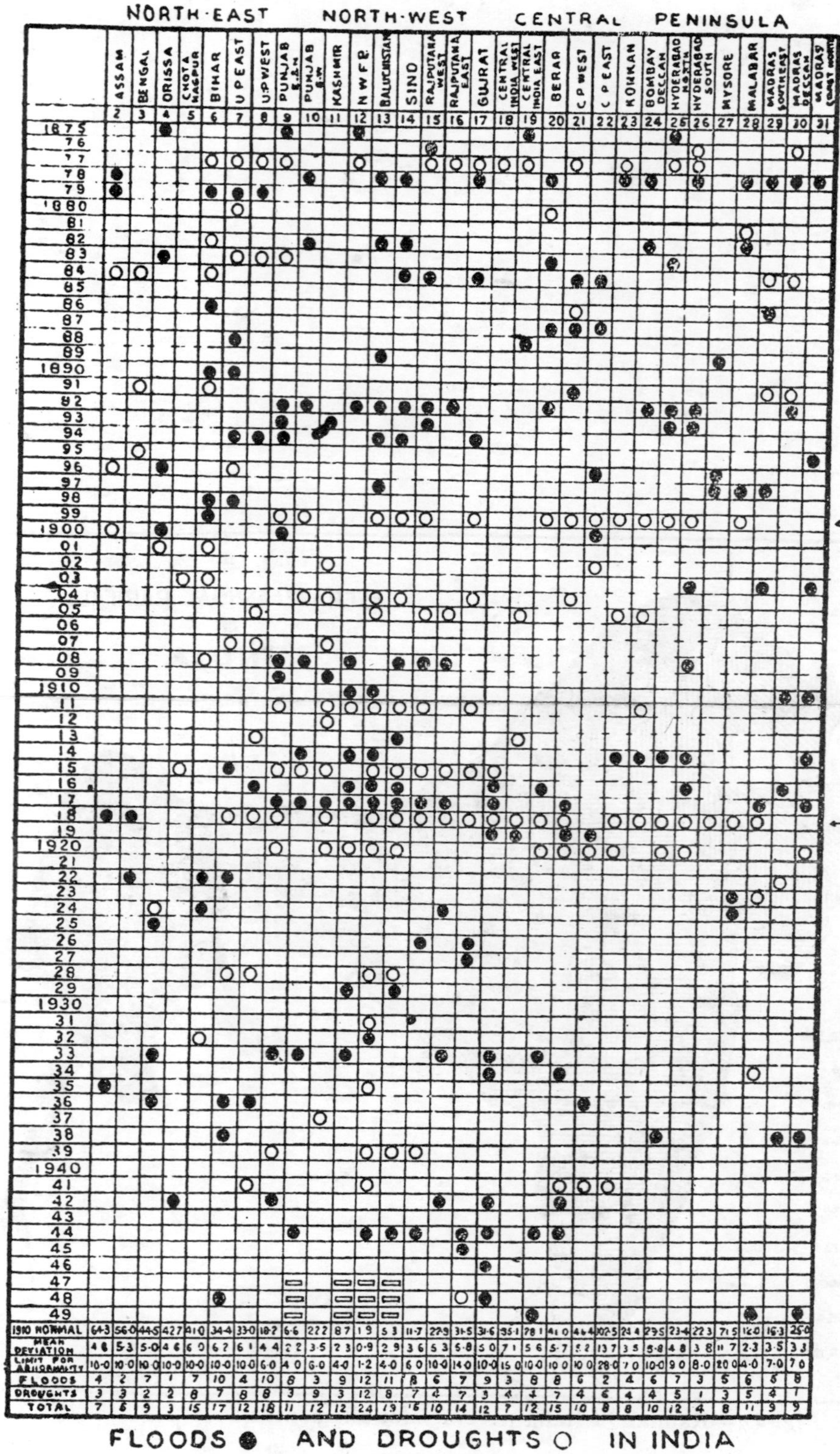

FLOODS ● AND DROUGHTS ○ IN INDIA

The famine years of 1877, 1899 and 1918 are conspicuous.

▭—Data not available—

Fig. 50. The incidence of floods and droughts in different parts of India during the period 1875—1949 (After Ramdas)

The ground flora vegetation also has a luxuriant growth in the season. Some areas get inundated and waterlogged which may cause great damage to tree seedlings by creating a deficiency of dissolved oxygen in the soil and its accompanying phenomenon.

(iv) Season of Retreating Monsoon.

The intensity of monsoon rain decreases considerably during the month of October and skies become clear and temperature begin to rise. The maximum temperatures during this month are remarkably uniform throughout India, between 80° to 90° F. The season is rather unhealthy due to stickiness and high humidity. During the next two months the temperatures begin to fall and cold sets in. Air movements over the sea and land, more or less, stop over large parts of India during October. Due to variations in temperature, local tropical cyclones develop in the region of Bay of Bengal and the east coast of Madras, and bring rain to these areas.

The vegetation conditions in deciduous forests attain their maturity and different colours of leaves show great contrasts in forests. The upper parts of the perennial ground vegetation wither away and much of the Bryophytic vegetation in the hills hibernates. Snow may fall at high elevations in the Himalayas. The characteristic feature of the Indian monsoon is its uncertainty. Large fluctuations from year to year may occur and sometimes rainfall is considerably delayed causing much damage by drought to field crops. The damage due to this delayed monsoon is greatly felt in the region, where the rainfall is the highest ; and the crop to suffer the most is rice.

The Climatic Regions of India

The most obvious and the simplest climatic regions of India are the rainfall divisions. They are given in Table 71. Climatic regions of India are shown in Figure 51. These may be described below:—

(i) Wet zone.—With rainfall more than 80″, comprising of the west coast with long and short dry seasons. The natural

TABLE 71

Climatic regions of India and their relation with vegetation, crops and other conditions.

Climatic region.	Natural region.	Example.	Natural vegetation.	Agricultural crops.	Dangers to vegetation and population.	Measures against losses.
1. More than 80" rainfall, (Wet zone).	(i) West coast.		Normally evergreen and semi-evergreen forest.	Rice.	Largely through floods rarely through droughts and earthquakes.	Against floods.
	(a) North-long dry season.	Bombay.				
	(b) South-short dry season.	Trivandrum.				
2. 40—80 inches rainfall (Intermediate zone).	(ii) Bengal and Assam. N. E. Plateau and middle Ganges.	Chittagong. Nagpur.	Deciduous sal forest. Most trees leafless hot weather.	Mixture of dry and wet zone in crops, rice predominates.	Drought may be the factor.	Irrigation.
3. 20—40 inches rainfall (Dry zone).	(i) Carnatic or Tamil, wettest months Nov. and December.	Madras.	Dry deciduous thorn forests and scrub.	Dry zone crops as millet, wheat.	Drought.	Irrigation.
	(ii) Southern and N.W. Deccan with mean Jan. temperature. 65 to 75 degrees F.	Hyderabad.	Deciduous forest only in better localities.	Wet zone crop (rice) grown with the help of irrigation and wells locally.		
	(iii) Upper Ganges plain with lower January temp, and higher July ones.	Delhi.	Deciduous forest.			
	(iv) North Punjab plain with winter rainfall	Lahore.				
4. Less than 20" rainfall (Arid zone).	(a) N. W. Lowland	Karachi.	Desert and semi-desert. Thorn forest and scrub	Permanent cultivation almost impossible without irrigation.	Drought.	Irrigation.
	(b) N. W. plateau.	Quetta				

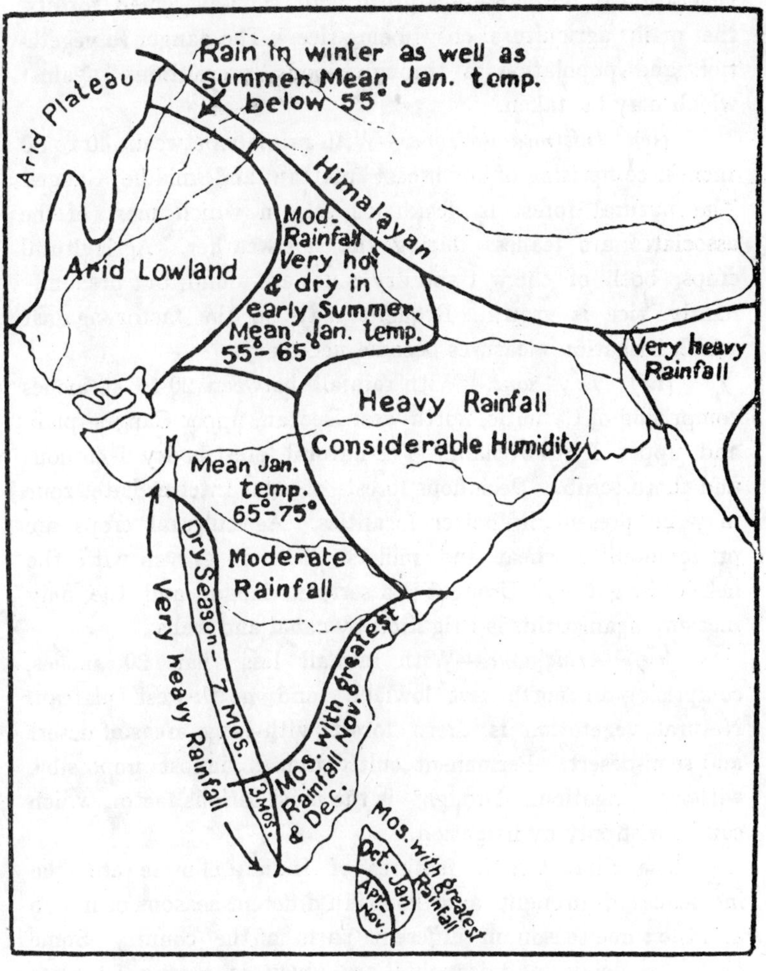

The major climate regions of India.
(after Kendrew, 1926.)

Fig. 51

vegetation is normally evergreen and semi-evergreen forests, the main agricultural crops being rice. The danger to vegetation and population is through floods, precautions against which may be taken.

(ii) Intermediate zone.—With rainfall between 40 to 80 inches, comprising of north-east plateau and middle Ganges. The natural forest is deciduous sal, in which most of the associates are leafless during the hot weather. Agricultural crops, both of the wet and dry zones are found, but predominantly rice is grown. Drought may be the factor against which irrigation measures become necessary.

(iii) Dry zone.—With rainfall between 20 to 40 inches comprising of Carnatic, north-west Deccan, upper Ganges plain and upper Punjab plain. The natural forest is dry deciduous and thorn scrub. Deciduous forest of the Intermediate zone may be present in better localities. Agricultural crops are predominantly wheat and millets. Rice is grown with the help of irrigation. Drought is a serious danger and the only measure against this is irrigation by canal and wells.

(iv) Arid zone.—With rainfall less than 20 inches, comprising of north-west lowland and north-west plateau. Natural vegetation is thorn forest with large areas of desert and semi-desert. Permanent cultivation is almost impossible without irrigation. Drought is the most serious factor, which can be met only by irrigation.

The characteristic features of Indian climate are the incidence of drought and floods in different seasons or in one and the same season in different parts of the country. Some data on floods and droughts are given in Figure 50 after Ramdas.

The climates of India in relation to vegetation and soils.

In 1937, Hosking prepared a map of India showing Meyer's ratio of precipitation to saturation deficit. This map though not quite satisfactory for showing an exact relationship between the development of soils and vegetation is the only one of its kind available to-day. When the first soil map of India was prepared in January 1943 by the Indian Council of Agricultural Research an attempt was made to show the relationship of the soils with the Meyer's ratio. In arid regions demarcated by Meyer's ratio of 5-25 desert soils belonging to alluvia were recognised. In the semi-arid regions with Meyer's ratio of 50-100 a number of soils designated by black clayey soils, brown sandy soils, red and yellow soils, kallar soils and part of desert soils were shown. In humid region between 150-250 Meyer's ratio black loam soils, black sandy soils, red and yellow soils and part of *usar* soils were shown. In perhumid region delimited by Meyer's ratio of 250-2000 dark reddish brown soils, coastal alluvium (laterite), red and yellow soils, dark reddish brown soils were recognised. Even in the 1949 map when a closer relationships between climatic conditions and regional soil types were presented the correlation between the climate and soil was not very satisfactory.

An attempt to relate Meyer's ratio of P/s.d. with Champion's preliminary forest types which are claimed to be climatic types, and "soil types" the following relationship seems to be evident:—

TABLE 72

Relationship of Meyer's ratio of P/s.d. to soil types and Champion's forest types.

Meyer's ratio.	"Soil types".	Champion's forest types.
5–25	Desert soils.	Tropical thorn forests.
25–50		Dry temperate forests.
50–100	Black clayey, brown sandy red and yellow, kallar and desert (in part) soils.	Tropical dry deciduous forests.
50–150		Temperate Himalayan forests.
150–250	Black loam, black sandy, red sandy, red and yellow and *usar* soils.	Tropical moist deciduous (sal) forests.
500–1000 and above.	Dark reddish brown, laterite and yellow, dark-red and brown soils.	Tropical wet evergreen forests.

The highest value of the Meyer's ratio in the tropical wet evergreen forests is 4000 found at Cherapoonji.

Although, a certain relationship between the Meyer's ratio and Champion's broader types and "Soil types" is obtained a close examination will, however, reveal important discrepancies. This is on account of the fact that Champion's terminology of tropical, temperate, subtropical, etc. and their subdivison is based on European standards, which necessarily are different from those found in India.

With a view to present a clear picture of comparisons and contrasts that Indian forest types (temperate, tropical, etc.) have with European and world types climatic pattern diagrams and climographs both for Indian and the world types (after Koppen) are given in Figures 52 and 53.

The differences between Koppen's temperate type and what Champion calls temperate forests are striking both in rainfall patterns as well as in climographs. Similarly, African and South American tropical types are very different from what Champion calls tropical from India. A closer study of the figures will reveal further contrasts which are not of degrees but are of kind. The principal disturbing factor which makes the European terminology inapplicable to India is the monsoonic nature of rainfall in the latter region. In this respect, India resembles China, Japan and other South-Asian countries, where rainfall has a similar monsoon pattern. Unfortunately, climatic and vegetational studies of this nature have not much progressed in these Asian countries and although, some work is done in Japan it is not easily available for comparison. The distribution of rainfall during different seasons in India and South-east Asian regions is given in Fig 54

The chief effect of monsoon pattern of rainfall on the soils is that periods of leaching and evaporation alternate, and soil development does not reach a maturity. On alluvia and other immature sites soils formed are largely azonal.

In coastal regions with a heavy rainfall and high temperature, that are peculiar to some parts of India laterite and lateritic soils are developed, which have no parallel in Europe.

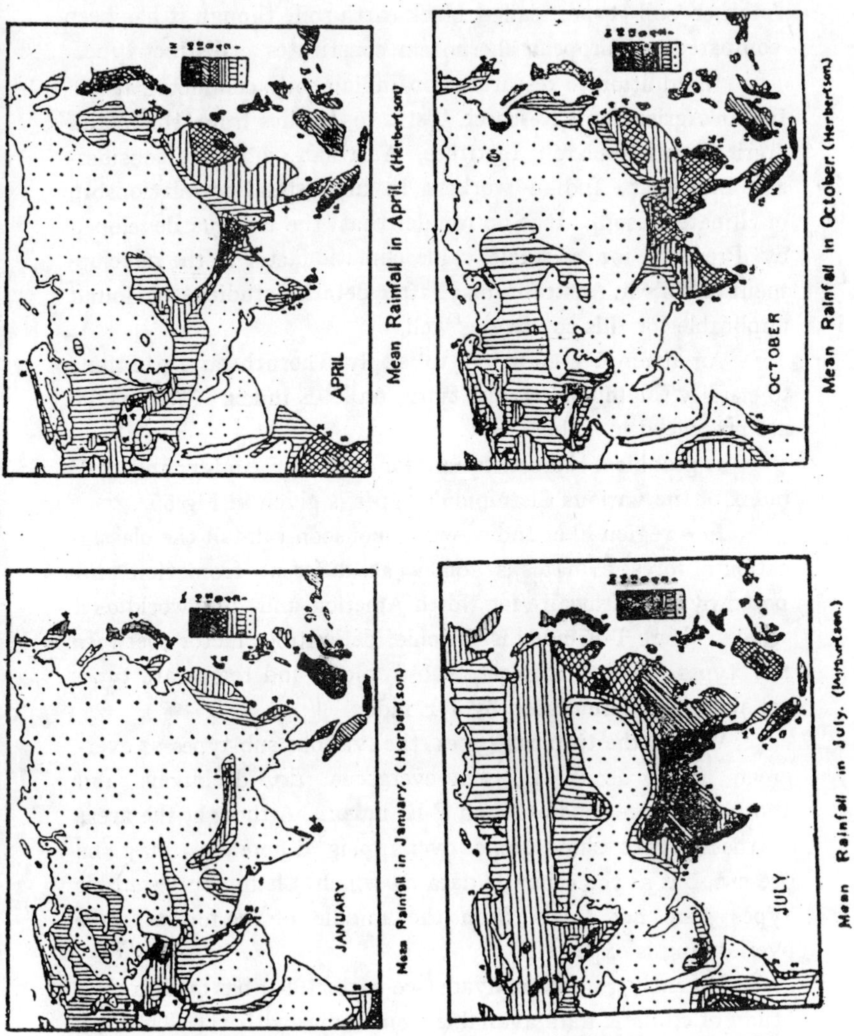

Fig. 54—The distribution of rainfall at different seasons in India and S. E. regions.

Another soil type, called black earth soil, though it has been compared to European chernozem constitutes a distinct type.

In addition to researches on Indian soils conducted at the Indian Agricultural Research Institute, studies from the Waite Agricultural Research Institute, Australia, will be of considerable interest to Indian workers in understanding relationship of climate to soil. It is no wonder that the formula developed by Prescott for a single value climatic factor in the development of soils in Australia may, after detailed studies, be found applicable for this country as well.

An attempt was made to apply Thornthwaite's formula to classify Champion's forest types on P-E index. The results are presented in Table 73.

A graphical representation for the P-E index and T-E index of the various Champion's types is given in Fig 55.

In a region like India with monsoon rainfall the classification of forest climates is somewhat different from that proposed by Thornthwaite for North America and the world as a whole. Here T-E index is the chief delimiting factor between the types. Thus, tropical, sub-tropical and temperate types are sharply differentiated by T-E index.

Within the tropical types the various sub-types as evergreen, moist deciduous, semi-evergreen, dry deciduous and thorn forest are separated on P-E index. Although, the areas of these types show some overlapping differences may be present, but as the climatic data on which Champion based his types were not taken from the middle of the forests, some overlapping is expected.

Recently, Subrahmanyam (see carter1954) has made a close study of climatic data available from India with the Thornthwaite's formula. He has been able to prepare maps showing moisture regime (Fig 56) and temperature regime (Fig57). This shows a correlation with forest types of India.

TABLE 73.

P/E and T/E ratios and indices of Campion's main types of forests, climatic data computed according to Thoruthwaite's formulas.

Forest type of Champion.	Typical station.		Jan.	Feb.	March	April	May	June	July	Aug.	Sept.	Oct.	Nov.	Dec.	
Tropical wet evergreen.	Akyab	P–E ratio	0.01	0.02	0.05	0.22	1.81	7.14	8.62	6.96	3.17	1.43	0.70	0.09	302.28
		T–E ,,	9.25	10.25	11.75	13.00	13.00	12.50	12.25	12.25	12.50	12.25	11.50	10.00	140.50
	Mangalore	P–E ratio	0.01	0.01	0.01	0.14	0.17	5.97	6.12	3.54	1.56	1.04	0.33	0.05	195.47
		T–E ,,	11.00	11.50	12.25	12.75	12.75	11.50	11.25	11.25	11.25	11.50	11.50	11.25	139.75
	Port Blair	P–E ratio	0.09	0.14	0.04	0.26	2.18	2.49	2.31	2.11	2.84	1.59	1.25	0.66	159.56
		T–E ,,	11.75	12.00	12.50	12.75	12.25	12.25	12.00	12.00	11.75	12.00	12.00	11.75	145.00
Tropical semi-evergreen.	Chittagong	P–E ratio	0.03	0.12	0.31	0.67	1.21	2.94	3.19	2.71	1.58	0.88	0.23	0.08	139.56
		T–E ,,	8.75	9.75	11.25	12.25	12.50	12.25	12.25	12.25	12.25	12.00	10.50	9.00	135.00
	Puri	P–E ,,	0.03	0.10	0.05	0.06	0.29	0.88	1.25	1.45	1.13	0.92	0.41	0.04	65.85
		T–E ,,	10.0	11.0	12. 0	12.75	13.25	13.25	13.00	12.75	13.00	12.75	11.25	9.75	115.05
Tropical moist deciduous forest.	Dehra Dun	P–E ratio	0.42	0.42	0.18	0.07	0.14	1.08	3.76	4.36	1.32	0.13	0.05	0.13	120.67
		T–E ,,	5.75	6.25	8.50	11.25	12.50	13.00	12.00	11.75	11.50	9.75	7.50	6.00	115.75
	Jalpaiguri	P–E ,,	0.04	0.09	0.17	0.45	1.49	3.35	4.56	3.55	2.76	0.62	0.02	0.01	170.90
		T–E ,,	7.50	8.00	10.00	11.75	12.00	12 50	12.50	12.50	12.50	11.50	9.75	8.00	128.50
	Seoni	P–E ,,	0.08	0.10	0.06	0.04	0.06	1.17	2.23	1.80	1.23	0.22	0.06	0.06	71.16
		T–E ,,	8.00	9.00	11.25	13.00	13.75	12.50	11.00	11.00	11.00	10.25	8.50	7.75	127.00
Tropical dry deciduous forest.	Rawalpindi	P–E ratio	0.54	9.43	0,37	0.24	0.12	0.20	0.84	0.97	0.39	0.07	0.06	0.19	44.31
		T–E ,,	4.25	5.00	7.25	9.75	12.25	14.25	13.75	13.00	12.00	0.50	6.25	4.50	111.75
	Jabbalpur	P–E ,,	0.11	0.10	0.06	0.02	0.04	0.90	2.62	2.35	1.04	0.20	0.06	0.04	75.52
		T–E ,,	7.50	8.50	11.00	13.25	14.50	13.50	11.75	11.50	11.75	10.50	8.50	7.25	129.50
	Cuddapah	P–E ,,	0.04	0.01	0.02	0.03	0.14	0.28	0.43	0.67	0.81	0.64	0.41	0.09	35.49
		T–E ,,	11.25	12.50	14.00	15.25	15.50	14.50	13.75	13.25	13.00	12.75	11.75	11.00	158.50
Sub-tropical wet hill and pine.	Shillong	P–E ratio	0.07	0.16	0.30	0.75	1.71	2.79	2.17	2.14	2.19	1.10	0.20	0.05	136.29
		T–E ,,	4.50	5.00	7.00	8.25	8.50	9.00	9.50	9.25	8.75	7.75	6.25	4.50	88.25
	Cherapunji	P–E ,,	0.12	0.39	2.29	6.04	9.29	20.36	20.65	15.64	8.74	3.00	0.30	0.03	868.42
		T–E ,,	5.00	5.50	7.00	8.25	8.50	9.00	9.00	9.00	9.25	8.50	7.25	5.50	91.75
Semi-tropical wet hill forests and pine.	Mercara (Coorg)	P–E ratio	0.02	0.04	0.07	0.35	0.84	5.09	8.63	5.04	1.96	1.47	9.49	0.09	240.84
		T–E ,,	8.00	8.75	9.50	10.00	9.50	8.50	8.00	8.00	8.00	8.50	8.25	8.00	103.00
	Pachmarhi	P–E ,,	0.10	0.10	0.05	0.04	0.06	1.29	4.16	3.96	2.26	0.29	0.10	0.08	124.91
		T–E ,,	6.50	7.50	10.00	11.75	12.75	11.50	9.75	9.50	9.50	8.75	7.00	6.00	110.50
	Ranikhet	P–E ,,	0.57	0.49	0.36	0.19	0.32	0.86	2.18	2.32	1.09	0.23	0.06	0.15	88.19
		T–E ,,	3.50	4.00	6.25	8.25	9.00	9.75	9.00	8.75	8.50	7.25	5.75	4.25	84.25
Himalayan moist temperate forests.	Chakrata	P–E ratio	0.62	0.71	0.68	0.20	0.26	1.28	4.18	4.45	1.41	0.16	0.07	0.24	142.47
		T–E ,,	2.50	2.75	4.75	7.00	8.00	8.75	8.00	8.00	7.75	6.50	4.75	3.50	72.25
	Simla	P–E ,,	0.70	0.76	0.58	0,39	0.43	1.17	3.18	3.34	1.05	0.16	0.09	0.22	120.74
		T–E ,,	2.25	2.25	2.20	6.50	8.00	8.75	8.00	7.75	7.25	6.00	4.25	3.25	68.75
Himalayan dry temperate forests.	Parachinar	P–E ratio	0.46	0.57	1.02	0.83	0.34	0.25	0.41	0.48	0.29	0.11	0.10	0.23	50 88
		T–E ,,	1.75	2.00	4.00	6.25	8.75	11.00	11.00	10.75	9.50	7.50	5.25	3.00	80.75
Temperate forests.	Darjeeling	P–E ratio	0.15	0.30	0.40	0.80	1.90	5.82	6.92	5.46	3.87	0.91	0.16	0.04	267.35
		T–E ,,	2.00	1.75	3.75	5.25	5.50	6.00	7.25	7.25	6.75	5.50	4.00	2.75	57.75
	Ootacamnad	P–E ,,	0.15	0.09	0.17	0.53	1.11	1.31	1.54	1.00	1.52	1.63	1.00	0.33	103.85
		T–E ,,	5.50	5.75	6.50	7.25	7.25	6.50	6.25	6.25	6.25	6.25	5.75	5.50	75.00
Tropical thorn forests.	Delhi	P–E ratio	0.15	0.07	0.08	0.03	0.06	0.33	0.99	0.80	0.51	0.05	0.01	0.05	31.37
		T–E ,,	6.75	7.50	10.50	13.00	14.25	15.25	13.75	13.50	13.00	11.75	9.00	7.00	135.25
	Jaipur	P–E ,,	0.05	0.04	0.04	0.02	0.05	0.25	1.01	0.90	0.36	0.03	0.01	0.04	27.94
		T–E ,,	7.25	8.00	11.00	13.25	14.50	14.75	13.00	12.50	12.75	11.50	9.25	7.50	135.25
	Poona	P–E ,,	0.01	0.01	0.01	0.05	0.14	0.68	1.00	0.48	0.67	0.49	0.12	0.02	36.70
		T–E ,,	10.0	11.0	12.75	13.50	13.25	11.75	10.75	10.75	10.75	11.50	10.75	10.00	117.85

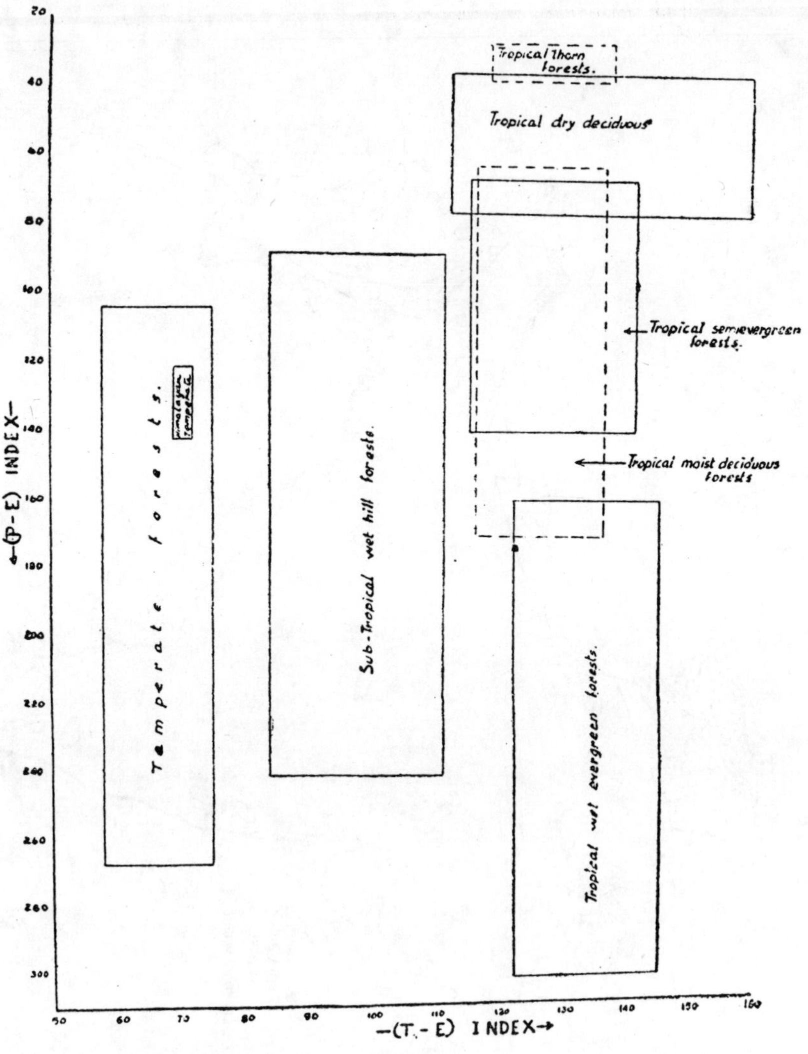

Fig. 52—The relationship of P-E and T-E index with forest types of India according to Champion.

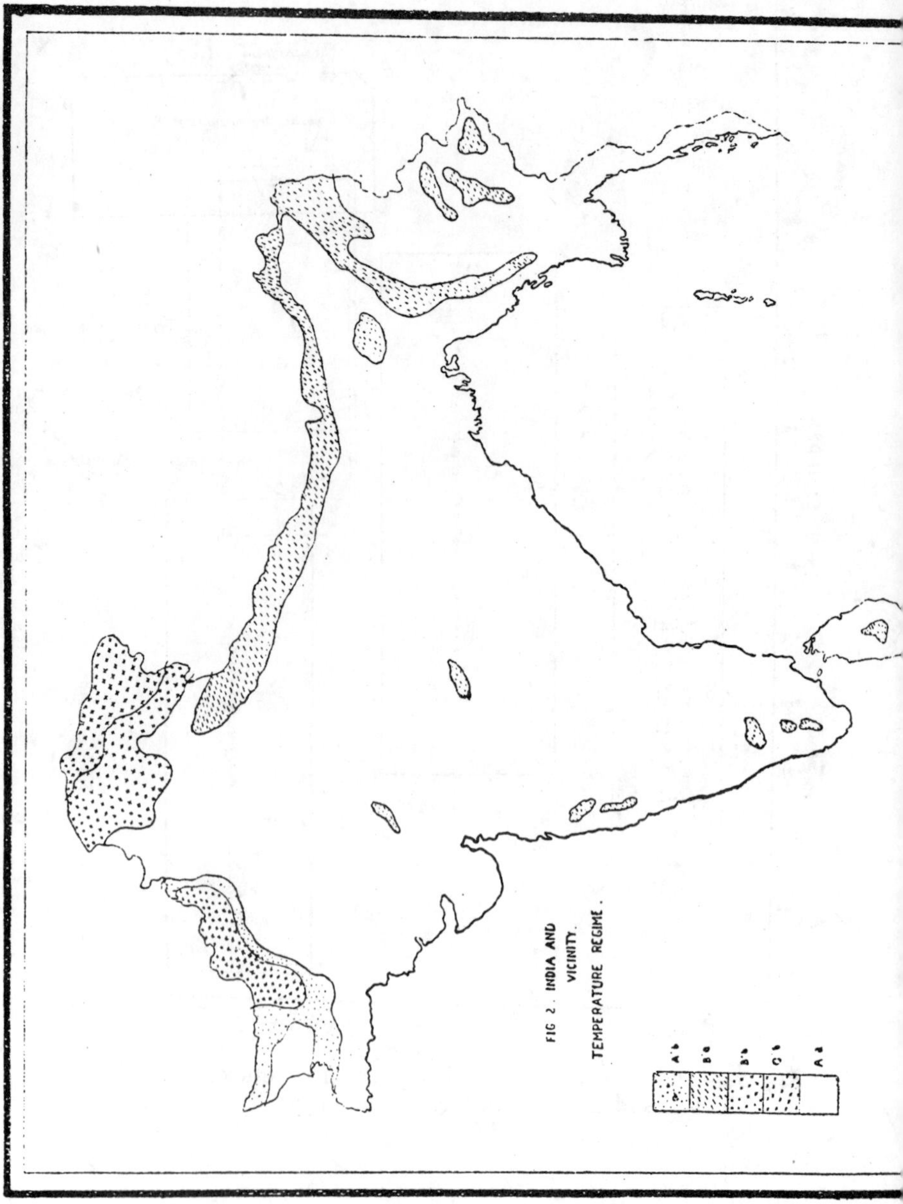

Fig 56—Temperature regime. A, B & C are Thornthwaite's terms for Megathermal Mesothermal and microthermal provinces.

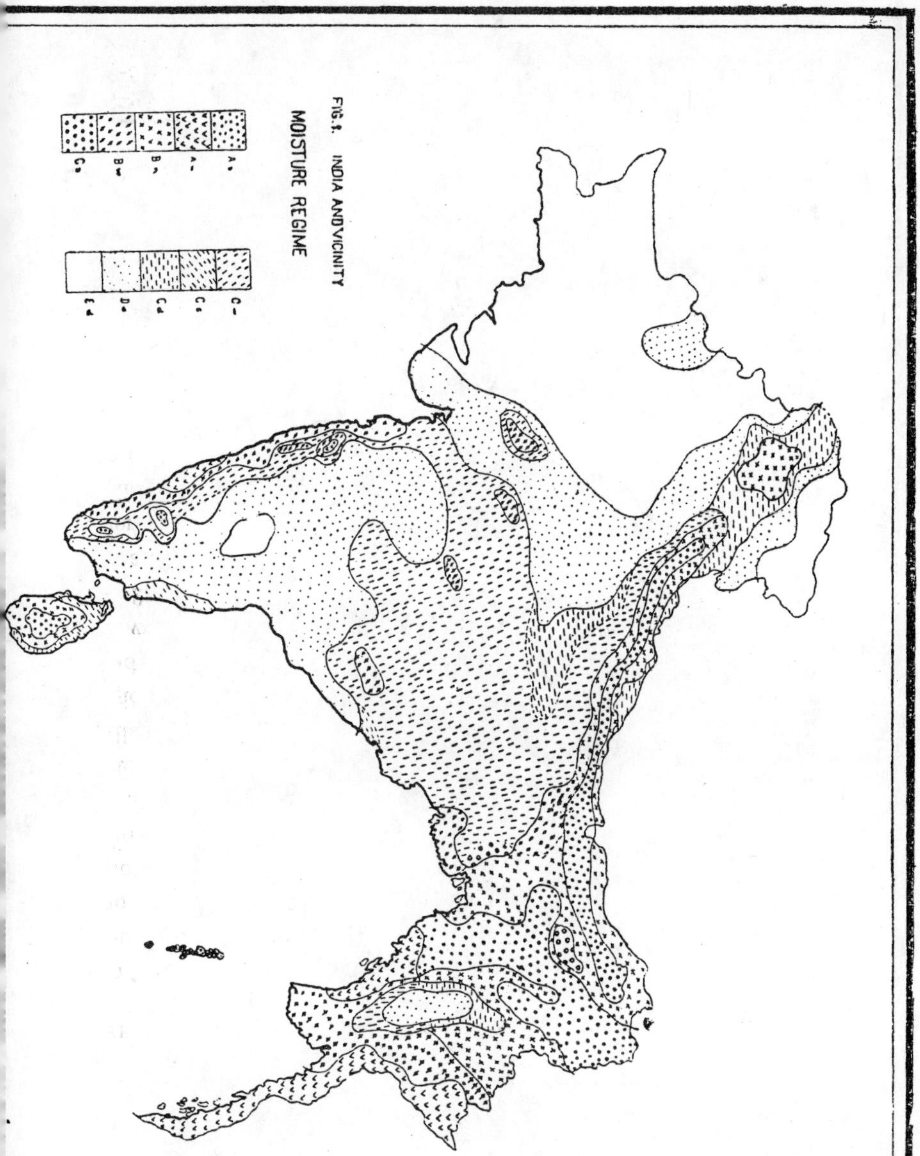

Fig. 57—Moisture regime. A. B. C. D. and E are Thornwaite's terms for Perhumid, humid, sub-humid, semi-arid and arid moisture provinces

15

The Light Factor

The importance of light in photosynthesis, metabolism and growth is well known. The study of light is of great ecological importance, since light plays an important part in biological equipment of species in colonisation and development of vegetation.

The amount of light reaching the surface of the earth and the length of day are functions of latitude and is greatly affected by the presence in the atmosphere of clouds, smoke, dust, gas molecules, water vapour, etc.

The solar radiation does not reach the earth in its entirety, but a great deal of it is absorbed in the atmosphere. The amount of light is greater at higher altitudes than at sea level. Lundegardh (1931) quotes Dorno to state that "from the total energy which the sun radiates upon the earth, only 75% reaches a height of 1800 metres, and according to the number of clouds, only between 24% and 50% reaches sea level. On the average, then, the land at sea level obtains through direct sun light only half the radiation recorded at 1800 meters".

All the light that reaches the earth is not utilised by plants but only red and blue parts of the spectrum are really useful in photo-synthesis.

The action of the different parts of the spectrum on living plants has been summarised by Lundegardh in Table 74.

TABLE 74

The action on plants of light of different colours and wave length.

Radiation.		Wave length (in mm).	Action on the plant.
Rontgen	...	0.00001-0.00018	Extremely harmful.
Ultra-violet	...	0.012-0.40	Very harmful in large quantity.
Violet Blue	... ...	0.40-0.49	Phototropism, morphological effects.
Green red	...	0.49-0.76	Photosynthesis.
Infra red	...	0.76—circa 600	Temperature effects.
Electric rays	...	2 mm to indefinite.	Not certainly known.

The ultra violet part of the spectrum is absorbed by plants more than yellow or red parts. Since the blue and red are of importance in photosynthesis, the differential absorption of light is of ecological importance. Water vapours in the atmosphere absorb greater amount of light of longer wave length, especially infra-red rays. During winter months higher percentage of red light and lower percentage of blue light reach the earth. On account of the high humidity, humid regions receive low hourly values of light than the arid regions and similar qualitative and quantitative differences are found between the arctic and the tropics.

Although variations occur, the amount of light received during the growing season on an average day in the tropics and in the temperate regions are not very different. However, slightly smaller amounts are received in the arctic than the tropics. In the arctic regions light is not a limiting factor for plant growth and since both arctic and desert plants have morphological devices against high light intensities it is believed that light is excessive of what is required for plant growth in the arctic regions. Shirley (1945) states that "light intensities in the tropics, though normally higher than in temperate regions, are not, in Veth's opinion, so much responsible for the luxuriant growth as the 12 month growing season. He states that growth increases rapidly upto 20% to 30% of full tropical sunlight, but decreases slightly as intensities increase above 50%. In tropical regions having a minimum of cloudiness, plants thrive on 1/5th to 1/12th of full light intensity. At greater intensities respiration due to the higher temperatures more than offset the benefits from increased light. This is the reason why many tropical plants thrive best in shade". At higher altitudes on account of the higher amount of ultra-violet rays, plant respiration is increased by 30-50%.

A minimum quantity of light is necessary for each species for its growth. In respiration the entire body of the plant is losing carbohydrates all the twenty four hours. This loss must be made up and a surplus built by photosynthesis confined to the leaves and that too within the limited hours of the day before any growth is possible. The minimum intensity of light which can just balance photosynthetic gains and respiratory losses is known as the compensation point. It is clear, therefore, that light intensity much above this point is needed for the plant. It is indeed this factor which limits the distribution of many plants in shade and in deeper regions of water.

Light Condition in Forests

A great deal of light intensity is absorbed by tree vegetation and light reaching the lower part of the canopy and for ground vegetation may be reduced by 90–99% of that in the open in some cases. In addition to the quantity, the variations in quality of light in a forest are also important from ecological standpoint. Leaves of most species generally absorb greater amounts of blue light than red and more of red than green.

Zederbauer (1908) measured the amount of light of different colours that is absorbed by leaves of different species. His results given in Table 75 show that the amount and quality of unabsorbed light under the canopy of different tree species on the same day are very different.

TABLE 75

The absorption of light by some forest trees.

(After Zederbauer, 1908).

Measurements on April 22, 1907 , at 10 O'clock, under a partly clouded sky.

Species.	Unabsorbed rays (in foot candles metre,					
	Red	Orange	Yellow	Green	Blue	Indigo	Violet
Abies alba.	75	100	200	250	300	500+	100
Pinus sylvestris.	225	320	500	500+	450–	500++	300

Measurements on June 14, 1907, between 3 and 4 p. m. under a cloudy sky at Mariabrunn.

	Red	Orange	Yellow	Green	Blue	Indigo	Violet
Picea excelsa.	2	7	12	100	40	200	100
Pinus sylvestris.	150	200	470	500+	500+	500++	200
Larix europaea.	50	80	90	250	200	500	100
Quercus pedunculata.	24	50	100	100	150	250	50

Since some parts of the spectrum are more effective for plant growth than the others the measurement of each portion of the spectrum is necessary to evaluate the ecological relationship of light with plant growth.

Discussing Zederbauer's results Zon 1917 stated "Zederbauers's investigations show in particular that the foliage of the so-called shade-enduring species, such as spruce, beech and fir, absorb most of the red, blue to violet rays, while the foliage of the light needing species, such as pine, larch and birch, absorb less of the red, blue to violet rays than the shade enduring species. In the shade of a forest stand composed of some shade-enduring species, the quality of the light must therefore be different from the kind of light that is to be found in the shade of a forest stand made up of light-needing species. Under the shade of a light-enduring pine forest, for instance, the light must still have a great amount of red, and particularly of blue and indigo, while in the shade of a shade-enduring spruce forest the light is poor in red, blue and indigo rays. Numerous laboratory observations similarly showed that the sun leaves and the shade leaves of the same species absorb different amounts of different rays of the spectrum."

According to Seybold and Eagle the forest floor species may be distinguished into three categories. Ratio af Chlorophyll A to B and ratio of Xanthophyll to Carotin in the 3 categories of plants in the forest are given in Table 76 The ratio of Chlorophyll A to B and of Xanthophyll to Carotin for immersed water plants are also included in the Table for comparison.

TABLE 76

The ratio of Chlorophyll and Xanthophyll and Carotin in plants growing in different conditions of light.

		Ratio of Chlorophyll A : B.	Ratio of Xanthophyll : Carotin.
Plants in forests.	Sun plants.	4.36	3.60
	Red green shade plants.	3.01	5.50
	Green shade plants.	2.60	5.30
Immersed water plants.	—	4.4	4.0
Submerged water plants.	—	2.3	5.7

A number of workers have measured light intensity in a forest and their results tend to show that under hardwood canopies during the leafless periods the light intensity is roughly 55% of that in the open. As leaf buds unfold the intensity of light beneath forest canopy decreases gradually.

Salisbury (1916) studied conditions of light in oak-Hornbeam woods of England. He recognised light phase in which trees are leafless and shade phase in which full leaves are present on the trees. For the growth of ground flora species in English woodlands Salisbury regarded that the amount of light intensity during the light phase is of the greatest importance.

The amount of light reaching the forest floor depends upon the height of the canopy, crown development of trees, age of the trees and phenological behaviour of the species.

Light intensities in some coppiced Oak-Hornbeam woodlands of South England are given in Table 77 after Salisbury.

TABLE 77

Light intensities in Quercus robur—Carpinus woods of Hertfordshire, England.
(After Salisbury, 1919).

Locality.	Age of coppice (years).	Shade-phase (%)	Shade-phase (%)
Stocking's wood. ...	7	50–60	0.9
,, ,, ...	11	37–48	0.16-0.76
Stocking's wood. (break in canopy). ...	11	—	5.4
Langley wood. ...	13	40.46	0.44–0.74
Pudler's wood. ...	14	29.1	—
Cutt's green. ...	4	61	1.3
Cutt's green (between stools).	4	—	5.5
Eight acre wood. ...	5	46–60	—
,, ,, ,,	6	41–55.5	0.78
Dowdell's wood. ...	17	38.4	0.83–1.1
Saucy wood. ...	10	42–50	0.33–1.2
Clapper's wood. ...	12	—	1.28
Secret Spring. ..	6	46	—
,, ,, ...	7	33.40	0.36
Bladder wood. ...	20	24	—
Milward's Park. (No undergrowth, open canopy). ...	—	60	—
Walshingham wood. ...	10	33-54	0.77
Walshingham wood (path). ...	—	50-100	—
Northaw Great Wood (No undergrowth dense canopy). ...	—	37-40	—
Total range. ...	—	61-24	5.5-0.16
Average. ...	—	41–48.2	0.8–1.4

Salisbury explained by differences in light intensity the absence of ground flora species in certain types of woodlands in which light intensity was not very different from those woodlands where ground vegetation was developed. He states that "the absence of ground flora from such woods is therefore probably due, in the case of the Pine and box, to the permanent character of the shade phase. In the case of the beech to the low intensity during the light phase and to the fact that the ground flora does not receive any periodic impetus through the coppicing of an undergrowth. In other words it is the intensity during the light phase combined with the time factor which determines the amount of the ground flora, the intensity during the shade phase being relatively unimportant".

The presence or absence of ground flora species in forests with less intensity of light is not really due to the light alone but as recent researches have revealed it is due to several other factors for example, temperature, moisture, plant growth, mineral nutrition, humus and root competition.

It is recognised that, although, light intensity during the shade phase seems to be lower, in actual conditions sun-flecks and movement in the leaf canopy of trees compensates for reduced light intensity and light may not be smaller in amount for the growth of ground flora vegetation. The measurements made by Oosting (1948) in a mixed Pine hardwood stand between 12 and 1 P.M. in 3 lines at various points at different times are shown in Table 78. These indicate the effect of sunflecks and the abnormally high records of light intensity at some places seem to compensate for the general low intensity.

TABLE 78

Light intensity in foot candles (with Weston Photometer) in a mixed pine-hardwood stand at the same spot at different times, showing differences due to sun-flecks.

(After Oosting, 1948).

Line 1		Line 2		Line 3	
300	500	500	400	100	100
300	500	500	300	200	400
200	300	300	400	300	100
200	300	400	200	200	600
300	300	300	200	200	200
300	200	300	100	100	200
200	200	400	3000	200	300
100	200	3600	2400	200	400
2400	200	4400	3600	300	300
300	200	400	200	100	200
300	200	2000	400	209	200
200	2400	400	400	100	200
500	200	400	300	100	200
500	200	600	400	100	100
500	200	1000	800	100	200
300	200	200	300	200	200
300	200	200	400	1600	200
4200	200	300	200	100	100
200	200	1200	5000	100	100
200	2100	1600	2400	200	100
200	600	300	4400	100	100
300	300	300	300	100	100
400	500	100	200	300	100
200	200	300	200	200	200
400	200	800	2000	200	100
Aver.532	432	832	1140	224	200

The effect of sun-fleck in tropical rain forests of Panama is shown in Table 79 after Allee (1926).

Here again the data seem to show that although light conditions in dense rain forests seem to be low, actually due to the effects produced by sun-flecks the total amount of light received by ground flora is not insignificant.

TABLE 79

Light variation in one spot through the day in tropical rain forest of Panama. The readings were taken in a palm thicket on top of a ridge.

(After Allee, 1916).

Hour.	Light intensity in foot candles.	Remarks.
8:05 ...	8.8 ...	Sun-flecks visible.
	9.6	
	11.1	
	11.6	
8:15 ...	10.3 ...	Some sun-flecks visible but not near. They cover, perhaps 1/20 of surface. Usually elongated but too narrow to cover the full width of test plate.
	10.8	
9:02	23.9	
	23.9	
	24.9	
	22.2	
	45.3	
	51.3 ...	Small sun-fleck 3 ft. away.
	58.1	
	59.8	
10:04 ...	25.3 ...	Sun-flecks not bright.
	24.9	
	27.4 ...	Sun-flecks brighter.
	27.4	
10:23 ...	24.9	
	24.4	
	23.9	
	23.5	
	24.3	
	24.3	
10:43 ...	432.4 ...	Sun-flecks now in centre of palm thicket. 5 ft. from test plate.
	901.0	
	1838.1	
	17.1 ...	Sun under cloud.
	16.5	
	23.3	
	25.9	
11:13 ...	19.1	
	18.2	
	1838.7 ...	Sun-flecks near.

Oosting states that "since light is obviously reduced under forest stand, it was once assumed rather generally that light is the controlling factor. Studies of trenched plots under forest stands gave results interpreted by many workers as indicating a greater significance for water since, within these plots, shade intolerent species for a time grew well when root competition for water and nutrients was eliminated by cutting off the roots of the dominant trees. Extensive investigations of conifer reproduction in the Lake States indicate that, for each light intensity, growth could be increased by reducing root competition and that at each level of root competition growth could be increased by increasing light".

The effect of root competition and shade on nursery transplants of white pine is indicated in Table 80. The data show that shade and root competition is more harmful than shade alone. Even in a high shade of over 4½ hours with no root competition 93% of the plants were alive but with root competition only 78% were alive.

TABLE 80

Influence of shade and root competition in a plantation of three year old white Pine transplants.

(After Kimball and Carter, 1913).

Low shade	High shade	With root	Competition	No root competition.	
		Number alive	Numbe dead	Percent alive	Percent alive
None.	None	49	58	46	56.5
	1 to 3 hrs.	11	12	48	81
	3 to 4½ ,,	10	8	56	84
	over 4½ ,,	25	7	78	93
Medium.	None	124	91	57	81
	1 to 3 hrs.	36	18	67-Av.alive	87-Av.alive
	3 to 4½ ,,	36	6	86 65%	90 85.5%
	over 4½ ,,	18	5	78	100
Heavy.	None	48	23	68	87
	1 to 3 hrs.	12	2	86	94
	3 to 4½ ,,	9	3	75	92
	over 4½ ,,	9	7	56	82

It is now considered that if other conditions of growth are favourable the amount of light intensity required by plants for survival is fairly low, between 1-5 per cent. The reason why under low conditions of light plants are unable to grow is that they have not produced sufficient amount of food reserve to enable them to bear the unfavourable effects produced by other factors.

Measurements of light intensity of different wavelengths for a few Indian forests are given in Tables 81 and 82 The differences according to the type of vegetation, their canopy and growth behaviour are quite obvious.

TABLE 81

Average light intensity in foot candles at 3½ feet level in some Indian forests

(From Puri, 1954).

Vegetation	Date and time of measurement	No. of observations.	Day light.	Red.	Green.	Yellow.	Blue.
Bamboos.	14-12-53. 11.45 A.M. to 1.40 P.M.	11	9.82	0.34	0.13	3.96	0,09
Evergreen.	19-12-53. 10.30 A.M. to 11.43 A.M.	10	45. 8	1.17	0.77	17. 3	0.36
Mixed sal at Barkot.	22-12-53. 11.30 A.M. to 12.45 P.M.	6	76. 0	1.43	0.97	25.53	0 52
	25-12-53.	5	76.96	2.32	1.11	31.04	0.62
Mixed miscellaneous.	9-12-53. 3-45 P.M. to 4-10 P.M.	5	25. 6	0.51	0.50	8.35	0.17
	10-12-53. 9-15 A.M. to 9.30 A.M.	3	74.13	2.08	1.01	28. 0	0.58
	27-12-53.	13	24.19	0.73	0.38	10.23	0.18
Sal forest at Timli.	1-12-53. 10.15 A.M. 2-12-53. 9.15 A.M. to 10.10 A.M. and 3.20 P.M. to 4.10 P.M.	19	61.14	1.34	0.87	19.64	0.54
Riverain deciduous.	15-12-53. 3 P.M. to 5-15 P.M.	8	100.60	2.46	1.50	32.31	0.70
	24-12-53.	4	57. 6	1.40	0.69	18. 8	0.37

TABLE 82

Average light intensity in food candles at 1 foot level in some Indian forests.

(From Puri, 1954).

Vegetation	Date and time of measurement	No. of observations	Daylight.	Red.	Green.	Yellow.	Blue.
Bamboos.	14-12-53 11.50 A,M. to 1.45 P.M.	11	12.66	0.34	0.12	4.15	0.09
Evergreen.	19-12-53 10.33 A.M. to 11.45 A.M.	13	32.42	0.88	0.54	13.22	0.26
Mixed sal at Barkot.	22-12-53 11.37 A.M. to 12.52 P.M.	6	56.67	1.56	0.90	22.0	0.56
	25-12 53	5	68.96	2.10	0 97	29.92	0.45
Mixed miscellaneous.	9-12-53 3.47 P.M. 4.12 P.M.	5	19.4	0.44	0.26	7.5	0.14
	10-12-53 9.17 A.M. to 9.35 A.M.	3	72.0	2.14	0.98	24.54	0.54
	27-12-53	13	25.09	0.61	0.30	9.28	0.16
Sal forest at Timli.	1-12-53 at 10.20 A.M. 2-12-53, 9.20A.M. to 10.15 A.M. & 3.25 P.M. to 4.12 P.M.	16	35.34	0.75	0.43	12.20	0.29
Riverain deciduous.	15-12-53 3.5 P.M. to 5.20 P.M.	8	96.60	2.40	1.64	32.20	0.73
	24-12-53	4	69.2	1.70	0.86	23.0	0 52

Light conditions in water.

Submerged aquatic plants are generally found to grow according to depth zones in a lake. However, very often this zonation is not solely on account of differences in the intensity of light reaching the plants at those depths as Pearsall (1920) and Misra (1938) have shown that the physical and the chemical qualities of the rooting medium vary according to depth. In shallower regions the lake bottom mud is coarse and deficient in nutrients whereas finer and richer particles of soil which can remain suspended in water for a longer time are carried to inner and deeper parts of the lake, where they settle down. Thus along with light intensity the character of the mud also changes at different depth.

Light penetration in bodies of water depends upon its turbidity, solute content, motion and planktonic overgrowth. Much of the light is absorbed by the upper layer of water. The blue and the green rays are reflected especially when the surface is rough thus imparting colour to the water. Even in the clearest water light reaching at a depth of 15 feet may be 50% of the surface intensity. At lower depths it becomes so feeble that photosynthesis cannot take place.

In considering the light relationship of plants the effect of other factors has unfortunately been not properly appreciated. The early methods of light measurements, especially those designed by Wiesner and Clements were photochemical, and tended to show only the physical effects of light. Photothermal method of Livingston brought to view the photo-thermal effects of light.

Of the total radiant energy falling on leaf not more than 5% may be used in photosynthesis and nearly all of the plant activities that require light rays do not utilize probably more than 10% of the heat value insolation. Thus, the remaining 90% of radiant energy is changed into heat, which is utilized in plant's physiological activities.

This shows that along with light, a good deal of heat is available to plants and the actual heating effect of this energy will be influenced by wind velocity. In an open area where air is not still, temperatures may not go very high even if the light intensity is high, but in a dense forest where air is usually still small amount of radiant energy may be sufficient to raise temperature of the plants. But in places where air movement is high and light intensity is low the unfavourable effect may be as much due to low light as low temperature and often to unfavourable factors accompanying low temperature conditions.

Light intensity determines C/N ratio of plant parts and this influences mineral nutrition. If carbohydrates produced are smaller or larger than the required amount the nitrogen metabolism is upset. Light also influences to some extent the absorption of iron, calcium and potassium in plants.

In forest stands, it is assumed that the death of a tree seedling is caused by the diminished light intensity. Some recent research has however, shown that tree seedlings which succumb in the forest on account of supposed low light intensity are not able to gain foot-hold even in grass-land communities where light conditions should be more than favourable. There has been some controversy over the question whether light or root competition is the factor responsible for the morality of tree seedlings in a forest, and the controversy is still not finally settled. Extensive researches on this point quoted by Shirely (1945) have shown that both the factors are responsible in producing mortality. He states "that at each light intensity growth could be increased by decreasing root competition, and at each level of root competition growth could be increased by increasing light. However, the increase in growth rate never was as great where only one factor was improved as where both were improved". The root competition is not only for water and nutrients but it is a root antagonism in which a plant is not able to thrive in a soil in which roots of certain other plants occur. This root antagonism due to production of toxic substances has been well illustrated in grass-land studies.

When plants are growing on poor soils the effect of low light intensity is greatly felt, but plants growing on rich soil without competition can withstand even smaller amounts of light intensity and grow well in shade. Thus Misra observed that *Peristrophe bicalyculata* grows in Uttar Pradesh on poor sandy soil in the open but on nitrogen rich loam in the shade.

Light and plant succession

Light is believed to be one of the important factors in determining the species that occur in an area and their successional trends in virgin conditions. The pioneer plants are usually considered to be light demanders, but as these develop the light intensity under the stand diminishes and invaders that come later are able to bear more and more shade.

Park (1931) has given ecological light units of four plant communities of the U.S.A. in Table 83.

TABLE 83
The ecological light unit ranges for some forest communities of the U.S.A.
(After Park, 1931).

		Ecological light units.
Pioneer cotton wood.	...	0.840 to 1.000
Pioneer conifer.	...	0.774 to 1.000
Pioneer black oak.	...	0.354 to 1.000
Climax sugar maple.	...	0.000 to 1.000

The light units of the pioneer cotton wood and conifer are greater than those communities that are climax in the area.

Among the pioneers that are considered to be light demanders in Indian forestry the two pines, sal, sissoo, *Acacia catechu* and even teak may be mentioned. Among the shade bearers are recognised deodar, silver fir, oaks and such other species. But this classification of Indian species is very arbitrary and there are no exact data for a scientific classification.

It has been shown clearly that light is not the only factor in determining the trend of succession in a forest but there are other factors of the locality which may be more important than light. Lundegardh (1931) has admirably discussed the light factor in plant growth and development and excellent summaries of Shirley (1935 and 1945) show the role of this factor in forestry. Burkholder (1936) has given a general discussion of the role of light in the life of plants and it is perhaps outside the scope of this work to enter into any detailed discussion of the role of light in photosynthesis, etc. Reference may perhaps be made to another important publication by Grasovsky (1929) in which he states that "the general conclusion reached is that the intensity and the quality of light reaching the forest floor are not the determining factors in accounting for the presence or absence of reproduction in the fully stocked forests where the investigation was made, and it appears from these studies that the absence of reproduction and presence of weakened growth under canopies is not necessarily due to diminished light intensity or to differences in light quality, but that other factors of the environment may be equally, if not more important."

Light conditions in evergreen forests.

Little work has been done in tropical forests and there are no data to show clearly the exact role of light in Indian forests.

MacLean (1919) has done considerable work on tropical rain forests of Brazil and some of his investigations may apply to Indian forests, therefore an account of these will be given here. A tropical forests, unlike European forest, does not have an unbroken canopy. The canopy in the former is broken up by trees in different strata. The normal light on the forest ground was nearly 1/140 of that in the open but in sun-flecks as high as 1/8 light intensity was recorded in tropical rain forest. The trees do not shed their leaves at one time of the year as in deciduous forests but some trees or the other in the canopy of a tropical forest keep on shedding their leaves and thus bring in more light intensity. Due to constant supply of sunlight, photosynthesis goes on unchecked throughout the year in the tropics. The leaves of tropical trees therefore do not show extreme variations as are found in sun and shade leaves in temperate regions. There is a perfect vertical distribution of vegetation in a tropical forest as a result of which light and other conditions do not become unfavourable and there is a luxuriant growth of vegetation due also to other factors of the environment being favourable.

McLean has measured light intensity in tropical forest of Brazil. In the white light the average intensities of three light are: Red, 0.883, Green, 0.850, Blue, 0.804. But the average intensities worked out in the leaf reflected spectra are: Red, 1.185; Green, 1.75; Blue, 1.0. This shows that in reflected light red decreases less than the other two colours. "On both counts therefore, red, not green, or blue, is preponderant in the diffuse light of the forest. That this circumstance renders the diffuse, reflected light of more value for photosynthesis than these might have been supposed from the photometer measures is evident."

Allee (1926) has given some very interesting results of the amount of light intensity in tropical forests of Panama. These are given in Table 83.

TABLE 84.

Maximum and minimum light intensities in shaded regions in the Panama rain forest between 8-00 a.m. and 4-00 p.m.

(After Allee, 1926)

Month and date.	Habitat.	% Sun-shine for the day.	Maximum light in ft. candles.	Hour.	Minimum light in ft. candles	Hour.
1-23	Shade from all trees only.	84	612	12-15	90	3-25
1-24	Ravine bed, Graph 5, fig. 13.	53	75	1-14	6	8-00
1-25	Palm thicket, Graph 3, fig. 13.	75	73	1-03	12	2-20
1-30	Palm thicket.	81	21	9-14	7	8-00
1-31	Palm lower vegetation removed.	100	377	11-88	66	9-45
2-5	On ground near spiked tree.	89	—	—	10	8-00
2-5	70-80 ft. up spiked tree.	89	631	10-22	173	10-20
2-7	On ground near spiked tree.	43	209	3-20	10	3-24
2-9	On ground near spiked tree.	77	36	2-30	8	10 04
2-9	75 ft. up spiked tree.	77	1,685	1-40	213	2-47
2-23	Shade of collecting bag on Ancon Hill.	70	1,369	10:12	468	4-20
2-27	Monkey cap palm, dense shade of sun-fleck types other habitats.	95	670	11-42	14	8-05
	a		18	11-41	2	8-30
	b		2	2-35	0.4	9-40
	c		8	12-00	4	9-50
	d		8	12-05	0.8	9-51
	Temperate forests in the United States.					
9-2	Beech and maple in Indiana, Graph 4, fig. 13.	—	39	11-55	19	3-00

A comparision between maximum sunlight and minimum light intensity in tropical and temperate forests in shadow is given in Table 35.

TABLE 85

Comparison of maximum sunlight measurements and minimum light measurements out of the shadow of nearby objects between 8-00 a.m. and 4-0 p.m. in tropical and temperate forests.

(After Allee, 1926).

(A) PANAMA (tropical rain forest).

Month and date	Habitat,	% Sun-light for the day	Maximum light in the candles	Hour	Maximum ilght in ft. candles	Hour
1-23	Small clearing, Barro Colorado.	84	6,024	12:00	—	—
1-25	Palm thicket, Sun-fleck, Barro Colorado.	75	4,000	10:30	326	1:31
1-30	Palm thicket, Sun-fleck, Barro Colorado.	81	20,541	8:30	681	8:10
1-31	Clearing thicket, Sun-fleck Barro Colorado.	100	12,325	11:10	1,220	9:45
2-5	Sun-fleck on ground near, spicked tree, Barro Colorado.	89	9,861	11:48	—	—
2-5	Sun-fleck, 70 ft. up tree.	89	8,216	8:43	2,737	9:42
2-7	Sun-fleck, on ground near spicked tree, Barro Colorado.	43	10,952	3:07	—	—
2-9	—do—	77	12,325	11:20	411	2:24
2-9	Sun-fleck, 75 ft. up tree, Barro Colorada.	77	20,747	1:40	4,520	1:45
2-16	On Gatun Dam. 75 ft. up tree, Barro Colorado.	17	4,019	2:16	1,067	4:00
2-23	Ancon Hill 75 ft. up tree,	70	10.957	2:24	1,917	3:35

(B) Temperate forests in the United States.

Month and date	Habitat,	% Sun-light for the day	Maximum light in the candles	Hour	Maximum ilght in ft. candles	Hour
7-12	Grassland Logan, Utah.	Cloudless	14,250	12-15	8,449	4-45
8-12	Grassland, Chicago.	,,	14,102	12:00	2,992	12:00 (Cloudly)
9-2	Sun-fleck, beech and maple woods, Bloomingdale Indiana.	,,	10,680	12-00	5,04	3-15 (Cloudly)

Some interesting data on the spectral composition of light in rain forests of Nigeria is given in Table 86.

TABLE 86

Spectral composition of the illumination in undergrowth in the rain forests of Southern Nigeria (Shasha Forest Reserve). Mean ratios as percentages, with fiducial values. Numbers in brackets below a mean indicate the number of ratios on which it is based.

(After Evans, 1939).

Stations.	Bright sunlight.			Sun completely obscured.		
	Blue filter	Green filter	Red filter	Blue filter	Green filter	Red filter
	A—Primary forest.					
III	7.05±0.49 (9)	5.94±0.18 (7)	30.0±1.5 (10)	9.02±0.27 (6)	6.16±0.17 (7)	20.1±1.4 (7)
IV	5.70±0.14 (8)	5.40±0.06 (5)	34.1±07 (6)	8.16±0.23 (14)	6.10±0.18 (14)	25.8±1.5 (18)
V	5.17±0.39 (27)	5.46±0.11 (24)	36.5±1.8 (29)	7.84±0.20 (13)	6.00±0.15 (16)	26.9±0.6 (13)
	B-Secondary forest.					
2	7.16±0.20 (3)	6.12±0.14 (4)	26.8±1.4 (5)	8.51±0.23 (13)	5.26±0.18 (11)	21.2±0.7 (12)
1	5.66±0.49 (9)	5.43±0.29 (6)	35.2±1.9 (6)	7.46±0.24 (10)	5 56±0.28 (9)	27.8±0.6 (11)
4	5.06±0.17 (14)	4.98±0.12 (16)	39.8±0.7 (17)	7.63±0.49 (7)	5.62±0.18 (9)	26.9±1.1 (10)
3	4.94±0.18 (22)	4.96±0.14 (19)	39.4±1.7 (20)	7.40±0.78 (6)	—	27.2±3.7 (4)

Light and Plant Growth

It is clear that a minimum intensity of light which should be above the compensation point, is necessary for plant growth. However, plants have specific light requirements and as such they are classified as (1) *Heliophytes* which grow best in sunlight and (2) *Sciophytes* which do well in shade. In leaves exposed to bright sunlight the chloroplasts tend to orientate vertically, chlorophyll is lesser in amount, anthocyanin development is in excess, palisade tissue may be more than one layer and cuticle is thicker.

The effect of photoperiod on the flowering of plants has already been discussed than in those that grow in shade. Germination of seeds is accelerated in light in some cases and is retarded in others. The role of light in germination is not fully understood. Bakshi (1952) found that the seeds of *Anisochelus eriocephalus* germinated better in light.

As far as the quality of light is concerned it is found that the red light or the higher wavelengths are conducive to cell elongation and etiolation. Violet and ultra-violet radiations on the other hand bring about dwarfening effects. In the alpine regions preponderance of shorter wavelengths, slow absorption of water at low temperature and desiccating winds all conspire to limit the growth of taller plants.

16

The temperature factor

The temperature of the atmosphere is controlled principally by the incoming radiation of the sun and is, therefore, a function of latitude, generally. However, altitude, distribution of land and water, wind direction, radiation and conduction of earth make a great deal of difference in determining ranges of temperature at the same latitude. On the whole, the average annual temperature of the northern hemisphere is slightly higher than that of the southern hemisphere and of the whole earth, as is seen from Table 87.

TABLE 87

Mean values of the temperature in deg. F. of both hemispheres and of the whole earth.

(After Haurwitz and Austin, 1944).

Temperature deg. F.

Part of the globe.	Mean yearly.	January.	July.	Annual range.
Northern hemisphere ...	59.3	46.5	72.3	25.8
Southern hemisphere ...	56.0	62.7	49.5	13.2
Whole earth. ...	57.7	54.6	60.8	6.2

Due to differences in land ond water surfaces in the two hemispheres each latitude in the northern hemisphere on a yearly average has a higher temperature than the corresponding one in the southern hemisphere. In tropical and sub-tropical regions the decrease of temperature and with latitude is very small but at higher latitudes in temperate and arctic regions temperature differences with latitude are larger. In the month of July, the southern hemisphere experiences winter and the northern summer; and the decrease in temperature from the equator towards the poles is much greater in winter than during the summers. As the temperature of the earth does not depend entirely on the incoming solar radiation, it is not a clear function of latitude. Much of the heat that falls on the earth during the day is radiated to the atmosphere during the night, however, some heat is stored and transmitted to lower layers of the soil. There are, therefore, diurnal and seasonal variations in temperature, besides the spatial variations. The annual range of temperature produces hot summers and cold winters, which are characteristic of continental areas. Over the oceans, oceanic regions and coastal areas, climates are equable.

The annual range of temperature is of great importance for the growth of plants, and plant life on the earth has been related to mean annual temperature zones. The plant classification systems devised by Schimper and other plant geographers have already been referred to. References have also been made on pages 62-69 to Raunkiaer's life form system of the plant classification in relation to plant's adaptations to temperature differences.

Temperature and altitude.

Temperature at high altitudes and in mountainous countries are governed by topographical features and air currents which may be formed by the mountain masses themselves. Throughout the toposphere, temperature decreases with altitude at the rate of 6° C per 1,000 metre in the lower part and 7°C per 1,000 metre in upper part. The average fall of 1°F of temperature for a rise of 300 feet in elevation is a mean value which rarely occurs. In nature, however, there is a temperature inversion; and valleys and lowlands are sometimes much cooler due to sinking in of the heavier cold air than the mountain tops. The low temperatures in depressions and higher temperatures on mountain tops may be due to other causes as well, such as radiation, wind, moisture, soil type, etc.

The amount of total radiant energy absorbed at higher altitudes is higher than at lower levels. At sea level 50% of total insolation penetrates the earth but at an altitude of 6,000 feet the amount of total insolation on a clear day may be as much as 75% of the total. The atmosphere of the earth contains liquid and solid impurities, which decrease rapidly upwards. At an altitude of 4,000 feet a larger part of these impurities are got rid off. At about 8,000 feet one is usually above half of the atmospheric vapour and suspended dust. The mass of the atmosphere also decreases at higher altitudes. on account of the thin atmosphere, ultra violet light of the spectrum is the most effective at higher altitudes. It is estimated that in England in a smoky area the intensity of the ultra-violet rays is reduced to about 1/3 of that in the open country. The ratio of the intensity of the ultra violet rays at altitudes of 250 feet., 5,000 feet. and 10,000 feet is 40:61:90 (Kendrew, 1938) respectively.

The amount of heat received in a day on a horizontal surface increases with an increase of altitude, as is shown in Table 88.

TABLE 88

Amount of heat in received in a day.

(After Haurwitz and Austin, 1944)

	Alt. ft.	Jan.	Apr.	July	Oct.	Year
Potsdam.	350	20	204	267	76	53200
Karlsruhe.	430	17	194	291	74	55170
Arosa.	6,200	86	276	350	176	81900

Although, greater amount of heat per unit area falls at higher altitudes strong winds tend to keep down temperature of the air above the earth. But if a thermometer wrapped in a good conductor is left on the surface of the earth it will register very high temperature as compared to the temperature of air above it. During the day rocks at higher altitude are heated and there is an intense evaporation from the soil. This is more so if the slope is facing the sun-side. The northerly and westerly aspects in the Himalayan hills are usually cooler and snow lines on these occur at altitudes lower than those on the southern and eastern aspects. During the night heat gained by rocks at high altitudes is dissipated and in a calm weather cold dense air sinks into the valleys lowering their temperature.

There are no absolute data for Indian regions to show the differences in temperature on North and South slopes, but some American data given in Table 89 show clearly the range of differences in maximum and minimum temperatures and their mean on SW and NE slopes, both during growth and rest periods in the forest.

TABLE 89

Summaries of maximum and minimum air temperature during 1912-16 on different slopes in forests of Idaho, U.S.A.

(After Larsen, 1940).

Datum.	Period.	Station 1 flat.	Station 2 SW slope.	Station 3 NE slope.
Averages of monthly maxima. ...	Growth.	73.5	4.5	770.0
...	Rest.	42.5	43.7	41.8
Averages of monthly minima. ...	Growth.	39.2	43.4	44.2
...	Rest.	22.0	25.5	25.9
Means. ...	Growth.	56.2	58.9	57.1
...	Rest.	32.1	34.9	32.9
Average annual range. ...	...	26 2	23.6	18.9
Absolute maxima. ...	...	97.0	99.0	95.0
Absolute minima. ...	...	—29.5	—20.5	—17.5
Extreme range. ...	...	126.5	119.5	112.5

The differences in temperature on the two slopes are probably due, in addition to the different solar radiation received, to differences in the amount of rainfall, snow fall, relative humidity and wind movement on the two slopes. Some interesting figures from U.S.A. are given in Table 90 for rainfall.

TABLE 90

Average precipitation for a 12 month period, on different slopes in forests of Idaho, U.S.A. (Based on five years data, (1912-1916).

(Aster Larsen, 1940)

		Total precipitation (inches)		
		Control flat.	S.W. slope.	N.E. slope.
		REST PERIOD		
October.	...	2.8	2.6	2.8
November.	...	5.6	5.3	5.7
December.	...	2.9	2.7	3.1
January.	...	4.2	3.9	4.2
February.	...	2.4	2.3	2.5
March.	...	2.6	2.4	2.7
April.	...	2.1	2.0	2.1
Average.	...	22.6	21.2	23.1
		GROWTH PERIOD		
May.	...	2.7	2.5	2.7
June.	...	2.6	2.5	2.7
July.	...	2.1	1.9	2.1
August.	...	1.0	0.9	1.1
September.	..	2.2	2.1	2.2
Average.	...	10.6	9.9	10.8
Year.	...	33.2	31.1	33.9

There is invariably a higher amount of rainfall on N. E. slopes than on S.W. slopes. In India the direction of the rain-bearing winds is from S.W. to N.E. and higher amount of rain should fall on the S.W. slopes. There is unfortunately no available data to show what exactly happens in these regions.

The data for snowfall on the N.E. and S.W. slopes for the Idaho forests are given in table 91.

TABLE 91

Average depth of snow at the three stations for the five year period, 1912-1916, on different slopes of Idaho, U.S.A.

(After Larsen, 1940).

	Depth of snow on the ground in inches.		
	Flat, Larch-Douglas fir.	SW slope, yellow pine.	NE slope.
October 15 ...	0	0	0
October 31 ...	0	0	0
November 15 ...	1.5	1.5	2.2
November 30 ...	8.4	5.5	8.8
December 15 ...	9.3	5.5	9.4
December 31 ...	18.1	9.2	19.1
January 15 ...	25.1	18.5	27.0
January 31 ...	27.3	16.9	27.8
February 15 ...	26.8	16.5	30.1
February 28 ...	24.2	6.1	24.8
March 15 ...	18.2	0.3	21.4
March 31 ...	7.2	0	12.7
April 15 ...	0	0	1.0
April 31 ...	0	0	0
Average ...	11.8	5.7	13.2

TABLE 92

	Length of snow cover (Days)		
	1	2	3
Average	128	107	141
Maximum	147	133	154
Average date of Disapperance of Snow Cover			
	March 24	March 8	April 8

There is not only higher amount of snowfall on NE slopes than the S W but snow lasts longer on the former than on the latter. As a result of higher rainfall and higher snow on NE slopes there is usually a higher relative humidity on these then on the SW. This is shown by data in Table 93.

TABLE 93

Relative humidity in percent on different slopes in forests of Idaho U.S.A (Readings taken at 6 P.M.).

(After Larsen, 1940).

	Average monthly.			Lowest monthly.			Absolute minimum.		
	Control flat.	SW slope.	NE slope.	Control flat.	SW slope.	NE slope.	Control flat.	SW slope.	NE slope
	REST PERIOD								
October.	84.5	74.1	73.5	75.3	65.2	65.1	45.0	35.0	35.0
November.	93.5	88.8	89.7	91.2	84.9	87.6	73.0	41.0	53.0
December.	90.8	89.5	90.0	80.9	87.7	84.2	76.0	43.0	43.0
January.	91.7	85.0	90.0	83.8	69.4	79.8	72.0	23.0	46.0
February.	82.0	74.0	80.9	79.0	64.0	74.3	40.0	25.0	37.0
March.	64.9	59.3	67.7	56.5	49.7	61.3	19.0	11.0	19.0
April.	51.9	52.5	60.0	46.4	46.0	47.6	21.0	25.0	19.0
Average.	76.9	74.7	78.8	73.3	66.7	71.4	...	...	...
Lowest.	...	...	...	...	...	...	19.0	11.0	19.0
	GROWTH PERIOD								
May.	52.8	53.6	59.2	44.7	43.6	50.1	14.0	12.0	21.0
June.	50.8	52.9	60.0	42.8	49.7	57.5	13.0	12.0	19.0
July.	46.5	46.0	54.9	36.0	32.3	43.1	19.0	20.0	25.0
August.	44.1	38.9	47.1	26.2	23.6	33.0	15.0	14.0	18.0
September.	66.2	54.6	60.4	60.8	45.6	56.0	17.0	12.0	18.0
Average.	52.1	49.2	56.3	42.1	39.0	47.9	15.6	14.2	20.2
Year.	68.3	64.1	69.4	57.8	55.1	61.8	...	...	...
Lowest.	...	...	...	...	...	...	15.0	12.0	18.0

The data for wind movement on NE and SW slopes are given for some American forests in Table 94. SW slopes, both during rest as well as growth periods, are more windy. NE slopes have much less wind movement.

TABLE 94

Wind movement in miles per hour on different slopes in forests of Idaho, U.S.A.

(After Larsen, 1940)

	Average.			Average maximum.			Absolute maximum.		
	Cont-rol flat.	S.W. slope.	N.E. slope.	Cont-rol flat.	S.W. slope.	N.E. slope.	Cont-rol flat.	S W. slope.	N.E. slope.
REST PERIOD									
October.	1.3	2.5	0.7	3.0	3.5	1.6	4.0	9.0	2.5
November.	1.3	3.0	0.6	3.4	3.8	1.5	4.2	9.2	1.9
December.	1.0	2.2	0.5	2.5	2.3	1.3	4.4	10.6	1.7
January.	1.4	2.7	0.6	4.7	8.4	1.7	11.2	14.0	2.5
February.	1.5	2.5	0.7	3.4	5.5	1.6	5.4	8.4	1.7
March.	2.0	3.5	1.0	4.7	9.2	2.7	7.0	11.9	4.3
April.	2.0	3.4	1.0	3.5	7.4	2.0	3.7	9.2	2.7
Average.	1.5	2.8	0.7	3.6	6.5	1.8	5.7	10.2	2.5
GROWTH PERIOD									
May.	2.0	3.3	1.1	4.2	7.2	2.5	5.7	8.7	2.9
June.	1.8	3.2	0.9	2.9	5.9	1.5	3.3	6.8	2.1
July.	1.7	2.9	0.9	2.9	6.0	1.5	3.7	7.6	2.0
August.	1.6	2.7	0.8	2.8	6.0	1.5	3.6	7.6	1.9
September.	1.4	2.6	0.7	2.8	5.8	1.4	4.0	8.6	2.0
Average.	1.7	2.9	0.9	3.1	6.2	1.7	4.0	3.7	2.2
Year.	—	—	—	—	—	—	—	—	—

There are better light conditions on SW slopes. On NW slopes, light intensity is almost half of what is present on SE slopes, as is seen in Table 95

TABLE 95

Light readings on different slopes in forests of Idaho, U.S.A.

(After Larsen, 1940)

Date.	Station 1 Flat.		Station 2 SW slope.	Station 3 NEslope.
May 2.	1.00		0.89	0.44
July 5.	1.00		0.70	0.25
August 27.	1.00		1.00	0.47
Standard.	1.00	Aver.	0.86	0.38

The growth conditions on the two slopes, compared with the control flat country are given in Table 96 and Table 97.

TABLE 96

Planted stock of 3-0 and 1-2 Western yellow pine on steep phase southerly aspects of different slopes in the forests of Idaho, U.S.A.

(After Larsen, 1940).

Location.	Station	Class.	Survival in percent.		Height in inches.	
			1915	1921	1915	1921
SW slope.	2 a	3-0	35	22	9.1	21
S slope.	2 b	3-0	45	24	10.0	23
SE slope.	2 c	3-0	26	12	11.0	24
SW slope.	2 a	1-2	61	27	6.7	18
S slope.	2 b	1-2	45	26	7.7	16
SE slope.	2 c	1-2	49	14	8.6	21

TABLE 97

Growth of the dominant trees in the plantations to the end of 1932 on different slopes in the forests of Idaho, U.S.A.

(After Larsen, 1940).

Series.	Location.	Species.	Average height in inches.	Number of trees.	Total age years.	Average height growth in inches.
A	Flat.	Western yellow pine.	182.4	51	20	9.1
A	SW slope.	Western yellow pine.	80.1	21	20	4.1
B	Flat.	Western yellow pine.	114.2	100	18	6.3
B	SW slope.	Western yellow pine.	88.2	36	18	4.9
C	Flat.	Western white pine.	75.5	12	17	4.4
C	NE slope.	Western white pine.	85.7	100	17	5.0
D	SW slope.	Western yellow pine.	74.4	12	17	4.3

The data are not complete enough to permit any conclusions on the effect of aspect on the growth of various tree species. But there is a general belief, not tested by exact and reproducible observations, that in the Himalayas all northern slopes carry a better plant cover than the southern; and the growth of plants is better on the former than on the latter.

Topography and temperature.

There are two main types of topography in mountains, the convex type and the concave type. The convex type has little control on the air above it and usually the temperatures do not become extreme. This may be compared to oceanic conditions of climate. The concave topography has a great influence on the air above it and it can be compared to continental condition of climate in producing extremes of temperatures. The concave valleys have a great deal of fog but valleys at very high altitudes enjoy bright sunshine and clear weathers. At Leh in Tibet at an altitude of 11,500 feet above sea level it is mentioned that by merely exposing a small blackened bottle containing water enclosed and shielded from air by a larger bottle of transparent glass, water could be heated to a high temperature.

Temperature and plant distribution.

In the Himalayas, Alps and other mountain chains as temperature changes are usually associated with altitude, there is a general zonation of vegetation from lower to higher altitudes. Thus, in the Himalayas there occur zones of subtropical to temperate and alpine types of vegetation, from lower to higher altitudes. The limits of these types are not clearly marked and due to differences in topographpy, soil and geology, sometimes there is an inversion of vegetation types.

On account of the difference in snow line in the two parts of the Himalayas, namely eastern and western one and the same type of vegetation occupies slightly different altitudinal zones. Taking single species the growth of trees shows an interesting relationship with altitude. As shown in Table 98, diameter of *Abies pectinata* trees of the same age class shows a negative correlation with the increase in altitude above sea level.

TABLE 98

The relation between altitude and growth in Abies pectinata

Altitude.	Diameter.	
	90 year.	100 years.
200–400	40 cm.	42
400–600	41 cm.	44
600–800	39	43
800–1000	35	38 cm.

In the Himalayas the growth of deodar shows somewhat similar effect of altitude (see Table 99).

TABLE 99

Distribution of sample plots of Cedrus deodara in India according to altitude.

(After Champion and Mahendru, 1933).

Altitude in ft. above sea level of the plots.	Site quality class Top height in feet at 120 years.				Total No. of plots studied
	1	2	3	4	
	140–120	120–100	100–80	80–60	
	Number of plots in each quality class.				
4,000— 4,500	1	—	—	—	1
4,500— 5,000	2	1	—	—	3
5,000— 5,500	2	1	1	—	4
5,500— 6,000	10	1	1	—	13
6,000— 6,500	7	10	2	—	19
6,500— 7,000	11	7	—	—	18
7,000— 7,500	18	9	1	1	29
7,500— 8,000	19	11	5	—	35
8,000— 8,500	3	5	1	1	10
1,500— 9,000	1	4	2	—	7
9,000— 9,500	—	2	1	2	5
9,500—10,000	1	1	1	1	3
Total:	74	54	14	5	147

The number of sample plots of better quality classes of deodar increases with decrease in altitude. Deodar has an intermediate altitudinal zone in the Himalayas and with increase of altitude above 8,000 ft. the plots of 1 and 2 quality classes greatly decrease with increasing elevation.

Similar correlation between altitude and quality occurs in plantations of *Tectona grandis* in South India and Burma.

As shown in Table 100 the quality of teak seems to deteriorate somewhat with a rise in elevation.

TABLE 100

Distribution of sample plots of Tectona grandis plantations in India and Burma by altitude and site quality.

(After Laurie and Sant Ram, 1940).

Elevation in ft. above sea level of the plots.	Site quality. I	II	III	IV	V	Total number of plots studied.
	Number of plots in each quality class.					
Up to 300 ...	6	63	75	13	2	159
300 - 600 ...	7	26	28	9	1	71
600 - 900 ...	5	29	12	2	2	50
900 -1,200 ...	—	2	10	15	2	29
1,200-1,500 ...	—	—	1	6	2	9
1,500-1,800 ...	—	1	1	2	4	8
Total	18	121	127	47	13	326

This relation between altitude and quality classes of deodar and teak may be due to causes other than the altitude as well. It has been seen in States of Madras and Bombay that depth of the alluvium on which teak grows best, decreases usually with a rise in altitude. On mountain tops, alluvium is generally very thin, and outcrops of laterite—a rock on which teak does not occur—come on the surface. Similarly, in the Himalayas, deodar occurs on flood plain deposits, which in Kulu do not ascend much higher up the mountains, so good growth of deodar is usually found mainly in valley bottoms on older alluvia.

In the mountains of India, especially in the outer ranges of the Himalayas, rainfall usually increases with altitude. So a correlation between altitude and quality class of some of the Himalayan forest trees may also be interpreted as with rainfall and quality classes.

In the Peninsular India, correlation between rainfall and altitude is not much significant, as the rainfall is brought about by monsoon winds arising from south-east. However, a somewhat insignificant relationship between rainfall and teak quality classes is shown in Table 101.

TABLE 101

Distribution of sample plots of Tectona grandis plantations in India and Burma by altitude and site quality.

(After Laurie and Sant Ram, 1940).

Annual rainfall in inches of plantation areas.	Site qualities. I	II	III	IV	V	Total
	Numbers of plots in each quality classes.					
40— 60	4	58	41	31	11	145
60— 80	1	12	13	5	—	31
80—100	3	22	34	2	—	61
100—120	6	17	24	8	2	57
120—140	—	—	—	1	—	1
140—160	—	2	—	—	—	2
160—180	—	—	—	—	—	—
180—200	4	4	2	—	—	10
200—220	—	—	—	—	—	—
220—240	—	6	13	—	--	19
Total	18	121	127	47	13	326

Most of the plots occur below a rainfall of 120 inches in a year. A few sample plots of good quality teak occur in regions of higher rainfall also showing that rainfall and altitude are not the only factors involved in the growth and development of this species.

At higher elevations, generally, temperature decreases and rainfall increases with increase in altitude and these factors greatly influence the development of soil and vegetation. Due to low temperature and high rainfall intensity organic matter of the soil increases at higher altitudes, and this is accompanied by increase in soil nitrogen, and decrease in pH.

The variation of soil organic matter and nitrogen with altitude and rainfall of some soils of the arid region of the State of Washington in the U. S. A. is shown in Table 102.

TABLE 102

Variation of soil organic matter with altitude in fine-textured soils of the arid regions of State of Washington, U. S. A.

(After Sievers and Haltz).

Altitude.	Annual precipitation.	Total No. in soil.	Carbon, Nitrogen, ratio.
feet	inches	per cent	
1,050	7.5	0.054	12.7
1,050	8.5	0.052	1.07
1,400	10.0	0.583	12.1
1,400	12.0	0.091	12.4
1,565	12.5	0.060	11.0
1,560	15.5	0.106	1.06
1,325	16.0	00.82	13.4
1,600	16.5	00.92	13.3
1,800	17.0	0.108	13.4
2,550	21.0	0.202	19.2

The data show a more or less increase in total N of the soil with increase in elevation. The C/N ratio of the soil shows a general increase only up to the altitude of 1,800 ft. and above that C/N again decreases, the causes of which are not known.

At altitudes above 5,000 ft. similar effect of altitude is seen on the soil nitrogen, both in grassland and forest communities. The data of Hockensmith and Tucker (1933) given in Tables 103 and 104 show this clearly.

TABLE 103

The relation of elevation to the nitrogen content of the soil under forest cover (U.S.A.).

(After Hockensmith and Tucker, 1933)

Elevation feet above sea level.	Percent of total nitrogen.		
	0 to 3 inches	3 to 6 inches	6 to 12 inches
5,350	0.341	0.136	0.018
5,590	0.311	0.107	0.061
5,750	0.251	0.135	0.067
5,900	0.388	0.132	0.059
6,200	0.421	0.162	0.061
6,300	0.310	0.159	0.062
6,400	0.786	0.317	0.051
6,810	0.426	0.123	0.053
7,300	0.329	0.132	0.047
7,440	0.401	0.130	0.052
7,560	0.231	0.140	0·061
8,200	0.744	0.111	0.041
8,250	0.277	0.046	0.027
8,635	0.351	0.155	0.051
8,850	0.511	0.068	0.042
9,000	0.442	0.100	0.020
9,050	0.798	0.031	0.032
9,150	0.787	0.109	0.038
9,200	1.151	0.096	0.029
9,250	0.901	0.071	0.031
9,750	0.971	0.044	0.024
9,800	0.822	0.056	0.028
10,450	0.898	0.079	0.051
10,500	0.990	0.082	0.050
10,700	1.483	0.121	0.046
10,800	1.982	0.135	0.062

In forest communities the surface soils show an interesting increase in nitrogen content from lower to higher altitudes, but in low layers of the soil increase in nitrogen content with altitude is not very proportional.

TABLE 104

The relation of elevation to the nitrogen content of grassland soils (U.S.A.).

(After Hockensmith and Tucker, 1933).

Elevation, feet above sea level.	Percent of total nitrogen.		
	0 to 3 inches	3 to 6 inches	to 12 inches
5,000	0.096	0.076	0.067
5,170	0.121	0.095	0.079
5,350	0.120	0.115	0.115
5,500	0.136	0.098	0.066
5,600	0.159	0.110	0.102
5,750	0.160	0.121	0.076
5,760	0.182	0.118	0.086
5,780	0.269	0.138	0.101
5,900	0.192	0.107	0.096
6,200	0.210	0.131	0.086
6,400	0.190	0.120	0.092
6,920	0.201	0.128	0.112
7,100	0.220	0.138	0.121
7,430	0.218	0.126	0.081
7,450	0.231	0.152	0.123
8,260	0.275	0.125	0.081
8,800	0.316	0.150	0.131
9,050	0.350	0.198	0.142
9,080	0.329	0.189	0.151
10,000	0.898	0.125	0.055
10,100	0.465	0.198	0.134
10,300	0.392	0.188	0.131
10,350	0.556	0.425	0.301
10,400	0.456	0.322	0.212
10,500	0.682	0.421	0.234
10,550	0.213	0.142	0.112

TABLE 104-Contd.

10,620	0.142	0.121	0.103
10,800	0.201	0.167	0.106
10,900	0.286	0.261	0.200
11,000	0.179	0.079	0.091
11,300	0.825	0.501	0.412
11,350	0.668	0.314	0.195
11,400	0.586	0.482	0.374
11,450	0.421	0.123	0.098
11,500	0.324	0.202	0.112
11,550	0.402	0.337	0.223
11,600	0.356	0.212	0.132
11,900	0.887	0.609	0.514
12.000	0.252	0.079	0.045
12,000	0.358	0.269	0.212
12,000	0.562	0.492	0.392
12,010	0.496	0.326	0.268
12,020	0.610	0.523	0.416
12,050	0.099	0.077	0.042
12,070	0.567	0.426	0.213
12,080	0.272	0.352	0.352
12,090	0.202	0.198	0.108
12,100	0.117	0.059	0.059
12,120	0.148	0.111	0.092
12,140	0.426	0.312	0.223
12,170	0.354	0.201	0.126
12,180	0.056	0.042	0.021
12,190	0.773	0.508	0.384
12,200	0.306	0.147	0.116

The data for Himalayan plant communities are given in Table 105 and 106. There seems to be a decrease in pH of the surface layer with altitude and increase in organic matter and nitrogen content.

TABLE 105

Relation between altitude and pH in conifer forests of the Kulu Himalayas.

(After Puri and Gupta, 1951).

pH Class.	—6.00	8.10 6.50	6.51 6.80	6.81 7.00	7.10 7.50	7.51 8.00	Total No. of soils studied for each community
	Number of soils for each pH class						
Spruce community (8,500 ft.	14	5	—	—	—	—	19
Silver fir community (9,000 ft.)	23	3	1	—	—	—	27
Blue pine community (4,500–,7000 ft.	2	12	5	3	—	—	22
Deodar community (5000—7,500 ft.	24	37	13	3	1	1	70
Chir pine community (4,000—5,000 ft.	—	8	8	2	5	—	23

TABLE 106

Relation between altitude, pH, loss on ignition, Ca and nitrogen in conifer forests of the Kulu Himalayas.

(After Puri and Gupta, 1951.)

Community.	Characters		pH		% loss on ignition		Ca in millequivalent				Soil profile type
	Altitude above sea level ft.	No. of profi-les	Mean for each com-munity	Mean ± S.E. for each soil type	Mean for each com-munity	Mean ± S.E. for each soil type	Mean for each com-munity	Mean ± S.E. for each soil type	Mean for each com-munity	Mean ± S.E. for each soil type	
Chir pine	4,000-5,000	2	6.90	6. 90±0. 27	17.35	17.35±9.20	40.80	40.80±12.85	0.42	0 425±0.719	Skeletal I.
Deodar	5,000-7,500	17	6.43		23.34		40.80		0.55		
Blue pine	4,500-7,000	5	5.56	6. 54±0.07	26.74	23.79±2.60	33.72	38.68± 3.63	0.51	0.45±0.48	Podsolized and sometimes brown earth type II.
Deodar-Blue pine	6,000	3	6.37		21.43		39 .07		0.36		
Spruce	8,500	4	5.85	5.88±0.12	25.70	30.15±4.11	48.30	62.71±5.75	0.71	0.785±0.76	Brown earth type III.
Silver fir	9,000	6	5.90			46.45	72.32		0.83		

Jenny has worked out a correlation between temperature, rainfall, altitude and organic matter to nitrogen of the soil.

A relationship between the percentage of life forms with altitude for Switzerland is given in Table 107 and relationship of rainfall with life forms for desert regions in the United States is given in Table 108.

TABLE 107

Altitudinal distribution of life-forms, Puschlav, Switzerland.

(After Smith, 1913).

Altitude.	No. of species	Percentage in life-form classes.									
		S	E	MM	M	N	Ch	H	G	HH	Th.
Above 2850 meters.	51	—	—	—	—	—	35	61	2	—	2
2550—2850 meters.	199	—	—	—	—	—	25	67	4	—	4
2250—2550 meters.	348	—	—	—	—	—	18	64	7	1	6
1900—2250 meters.	492	—	—	—	1	3	13	68	8	1	6
1550—1900 meters.	487	—	—	—	3	4	11	62	10	1	8
1200—1500 meters.	449	—	—	2.5	2.5	4	7	60	9	1	14
800—1200 meters.	604	—	0.2	2	3	5	15	55	9	2	19
Below 850 meters.	447	—	0.2	3	4	3	6	55	8	1	21
Normal spectrum.	400	1	3	6	17	20	9	27	3	1	13

TABLE 108

Life-form spectra for five desert regions in the United States.

(After Paulson 1915)

The District.	Rainfall	No. of species	P (N)	Ch	H	G	Th
Akron, Colorado (H. L. Ghautz).	47 cm	79	—	19	58	8	15
Toocle Valley, Utah, (Kearney, et. al.)	40 cm	116	2	23	46	3	14
Tucson, Arizona (V. M. Spadling).	27 cm	266	18	11	24		47
Salton, Sink, Arizona (V. M. Spadling).	20 cm	81	33	6	14		47
Death Valley, Calif. (Raunkiaer Cooille)	—	294	26	7	18	2	42
Normal spectrum.	—	400	43	9	27	3	13

On the whole, Therophytes are more common at lower altitudes and Chamaephytes occur more at higher altitudes. Phanerophytes are confined again to lower altitudes. Hemicryptophytes have a large percentage throughout, though at lower altitudes their percentage is somewhat lower than at higher altitudes.

Therophytes have a higher percentage in regions of lower rainfall. In regions of high rainfall Hemicryptophytes and Chamaephytes have higher percentages.

Love and Love (1949) have shown an interesting relationship between latitude, altitude, and number of monocots and dicots; the polyploid species of which seem to increase with altitude between Ca 17⁰ N—81⁰ S and with altitude from 2,500 Km to 449.000 Km. as is shown in Table 109.

The same authors have also shown a relationship between polyploids and various life forms in four regions. The data are given in Table 110.

TABLE 109

The frequency of polyploid angiosperms in different regions

(After A. Love and D. Love, 1949)

Area.		Latitude.	Altitude Km.	Number of angiosperms.	%Cytol. known	% of polyploids.		
						Monocot.	Dicot.	Total.
Timbuctoo	...	Ca.17° N. lat.	—	138	21.0	67.0	31.0	37.0
Cyclades	...	36½°—37½°	2.500	1184	48.3	44.7	31.2	34.1
Sicily	...	36½°—38½°	26.000	2346	37.6	51.5	32.8	37.0
Central Hungary	...	46° —49½°	93.000	2086	68.6	—	—	47.1
Schleswig-Hostein	...	54° —55°	16.000	1131	91.7	62.0	46.2	50.2
Denmark	...	54½°—58°	44.000	1306	90 4	70.2	47.4	53.5
Great Britain	...	50° —61°	311.000	1630	72.6	75.1	49.8	56.7
The Faeroes	...	Ca 62°	1.400	324	90.1	80.0	50.3	61.3
Iceland	...	63½°—66½°	106.000	440	88.2	84.0	52.5	63.8
Sweden	...	55½°—69°	440.000	1645	87.3	74.8	48.7	56.0
Kolgujew	...	69°	3.500	—	—	—	—	64.0
Finland	...	60°—70°	383.000	1285	85.5	77.2	48.9	57.3
Norway	...	58°—71°	323.000	1351	86.5	76.1	49.8	57.6
Southern Greenland	...	60°—71°	—	328	81.4	92.7	60.2	71.9
Spitzbergen	...	77°—81°	63.000	135	82.1	95.0	61.4	73.6

TABLE 110

Biological spectra and polyploidy in different regions.

(After A. Love & D. Love, 1949).

		Pardubice.			Sjaelland.			Pite Lappmark.			Sassen quarter.		
		No.	%	%polypl.	No.	%	%polypl.	No.	%	%polypl.	No.	%	%polypl.
	Monocot.	...	...	...	...	...	...	...	...	...	...	...	...
NM	Dicot.	61	9.2	47.5	62	8.7	51.6	19	9.4	47.4	1	1.8	100.0
MM	Total.	61	6.8	47.5	62	6.2	51.6	19	6.1	47.4	1	1.1	100.0
	Monocot.	...	...	...	...	...	...	...	...	...	...	...	...
Ch.	Dicot.	23	3.5	39.1	34	4 8	52.9	43	21.2	46.5	19	33.3	63.1
	Total.	23	2.6	39.1	34	3.5	52.9	43	13.9	46 5	19	21.1	63.1
	Monocot.	99	41.6	74.7	114	41.9	75.4	58	54.2	86.2	24	72.7	95.8
H	Dicot.	347	52.4	48.1	356	50.1	47.8	113	55.7	51.3	33	57.9	66.7
	Total.	446	49.6	34.0	470	47.8	54.5	171	55.3	63 2	57	63.3	76.7
	Monocot.	99	29.0	78.2	75	27.6	84.0	35	32.7	97.1	8	24.3	100.0
G	Dicot.	20	3.0	60.0	21	3.0	52.4	2	1.0	100.0	1	1.8	100.0
	Total.	89	9.0	74.2	96	9.7	77.1	37	11.9	97.3	9	10.0	100.0
	Monocot.	44	18.5	63.6	62	22.8	58.1	13	12.2	76.9	1	3.0	100.0
HH	Dicot.	27	4.1	55.6	35	4.9	54.3	7	3.4	57.1	...	...	...
	Total.	71	7.9	60.6	97	9.9	56.7	20	6.4	70.0	1	1.1	100.0
	Monocot.	26	10.9	34.6	21	7.7	42.9	1	0.9	100.0	...	...	...
Th	Dicot.	184	27.8	40.2	203	28.5	38.9	19	9.3	42.1	3	5.2	33.3
	Total.	210	23.2	39.5	224	22.9	39.3	20	6.4	45.0	3	3.4	33.3
Total		900	100 %	52.3%	983	100.%	53.2%	310	100.%	63.2%	90	100 %	76.7%
	Dicot.	662	73.6	46.2	711	72.3	46.3	203	65.5	49.8	57	63.3	64.9
	Monocot.	238	26.4	69.3	272	27.7	71.3	107	34.5	88.8	33	36.7	97.0

Temperature gradient in water.

Submerged aquatic plants are subject to lesser fluctuations of tempeature in their surrounding. Due to the cooling effect of evaporation surface water does not get readily heated up. The latent heat of water being high it does not get cooled readily in the night time. These effects will be more in larger bodies of water. Moreover, wind effects and convection currents mix up water and lessen the effects of local heating or cooling. Diurnal variations in temperature can be found, however, in shallow ponds.

In deep waters of the temperate Himalayan lakes thermal stratification may be found in the summer. At this time there are (1) vertical gradient of gradually decreasing temperature from the surface which is called the *epilimnion*, followed by (a) a short zone of rapidly falling temperature—the *thermocline* and (b) the bottom cold zone the *hypolimnion* in which there is no temperature gradient. As the winter approaches the surface layers of water loose heat to the atmosphere and the cooled water being heavier sinks down. This process gradually brings down the temperature of the whole lake, but before a uniform temperature is attained the entire body of water is churned up. A period of summer stagnation followed by water circulation leads to release of plant nutrients and their mixing up from the bottom mud into the lake water.

Temperature in relation to growth and distribution of plants.

As is well known higher temperatures lead to increased transpiration and excessive loss of water. A desiccation of plant tissues on the other hand may arise on account of slow absorption and ice formation at very low temperatures. The extremes of temperature within which the plant can grow is a specific phenomenon, as plants growing both in the arctic region and in hot springs are found.

Since the optimum temperature for respiration is higher than for photosynthesis the plants grow better at temperatures favourable to the latter process. Thus, for growth of most fo the plants in this country a temperature near about 30° C is excellent.

It is also known that for growth and development periodic exposure of plants to low temperature is beneficial. For instance in temperate regions the dormant buds of deciduous trees, will burst into activity following a cold spell. However, if such buds are kept in a warm house, they fail to open in the summer. The phenomenon of vernalisation, similarly is too well known to be described here.

Frost injury is a common phenomenon in northern parts of India. It happens due to sudden cooling of growing buds in winter nights. Several trees and wild plants are susceptible to it.

High temperature injury is seen following forest fires. Only such trees are capable of surviving which have thick bark or regenerative underground parts.

17

The Atmospheric Factor

The thick mantle of the atmosphere surrounding the earth is composed of about 79% nitrogen, 21% oxygen, 0.03% carbondioxide, traces of a few other gases, and water vapour in varying amounts. Smelter gases near industrial towns may also be present in significant quantities. Besides these gases solid particles like dust, smoke, micro-organisms, pollen grains, spores, etc. may remain suspended in the air. Any movement in the air or wind tends to mix up these elements of the atmosphere thus avoiding local concentration of the gases which otherwise diffuse slowly according to their individual concentration gradients.

The gases of the atmosphere also exist in the soil air, aerenchyma of plants and dissolved in fresh and sea waters and tissue fluids of organisms. They diffuse into and out of each of these systems depending upon a number of factors such as pressure, temperature, etc. They get chemically bound and released from a large number of natural compounds found both in the living and the non-living worlds. Each gas in fact passes through a regular cycle which tends to maintain balance in nature. Thus, oxygen is consumed by all living organisms for respiration and is replaced by the green tissues in light during photosynthesis. Carbondioxide is given out by organisms in respiration and is consumed in photosynthesis. Nitrogen, on the other hand, is fixed up by some micro-organisms living in the soil and plant tisssues and is returned to the atmosphere by denitrifying processes. Water vapour content of the atmosphere is largely controlled by precipitation, presence of liquid water and conditions for its evaporation through plants, soil and water surfaces.

The atmosphere which really matters to plant life directly is just that much as surrounds it immediately for gaseous exchange. Indirectly, however, the entire body of the atmosphere is responsible for controlling climate, temperature, light and other conditions already discussed in the preceding chapters. Soil aeration has also been dealt with and here only the shoot atmosphere will be taken into account.

Ordinarily oxygen concentration in the shoot atmosphere does not limit the growth and distribution of land plants. Even in a dense vegetation it may not fall below 10%. Such a low concentration may be reached during the night time near the soil surface but the gas is replaced by photosynthesis during the day. In case of aquatic plants, however, there seems to be a real difficulty in maintaining optimum pressure of oxygen as this gas is sparingly soluble in water and its rate of diffusion is too slow to replace its rapid depletion at lower depths. The system of large inter-communicating air chambers within the plants is a storage of gases especially of oxygen as the latter gas is produced in abundance on account of active photosynthesis during the day.

Carbondioxide concentration in the shoot atmosphere is more often a limiting factor in carbon-assimilation. Hence plants do better on soil rich in organic matter in which micro-organisms are actively producing the gas in respiration. Toxic concentration of CO_2 is hardly ever reached in the shoot atmosphere and it is only through the soil air and roots that carbondioxide injury to plants may be caused. Because this gas is readily soluble in water aquatic plants are better supplied with it than land plants.

Nitrogen in the shoot atmosphere is of no consequence to the plant as the gas is inert and can be fixed only by soil organisms from the soil air.

Water vapour in the air is responsible for its humidity, and its concentration at any time is largely controlled by temperature. At any temperature a maximum amount of vapour can be held by the air. Its actual pressure (absolute humidity) expressed as percentage of the maximum at that temperature is known as relative humidity and the difference between the two expresses the saturation deficit of the air. The saturation deficit is directly related with the evaporating power of the air and hence also controls to a great extent the rate of transpiration. Therefore, a study of the humidity is of utmost ecological importance. Since the air is capable ef holding lesser amount of moisture on çooling, precipitation in form of dew may occur with extreme diurnal variations in the temperature. Such conditions arise mostly in desert areas and low forms of vegetation are thereby benefited. In dense vegetation however, the atmosphere is kept moist by large scale transpiration, and any increase in humidity is thus not due to fall in temperature. An increased humidity is favourable to precipitation from the clouds. Hence rainfall is generally higher in forest areas.

Mosses and lichens are capable of absorbing moisture from the vapour present in the air. Hence they can be found on tree barks and stones more abundantly in humid surroundings.

The atmospheric air is not still. The causes and effects of wind have already been discussed under the climatic factor. Its direct effect upon the plant in increasing the rate of transpiration, causing mechanical damage and dissemination of pollen and plant propagules is very important. The velocity of wind at the surface increases in proportion to the extent of clear sweep that it obtains on flat plains, along sea coasts and at high altitudes on mountains. Vegetation and other obstacles work as barriers in reducing the velocity. A forest with staggered canopy of trees may reduce its velocity by 80% and act as a wind break. The wind drops its load of suspended particles at such breaks which give rise to sand dunes and loess. Deep shelter belts of *Prosopis juliflora* are being planted to arrest the eastward march of Rajputana Desert in India.

In the Indo Gangetic plains of India dust storms are frequent on accouut of high wind velocity playing on loose alluvia uncovered by vegetation. Isolated trees in orchards are uprooted and crop plants get lodged on this account. On the eastern sea coasts cyclones and gales uproot trees and kill windward buds due to salt sprays. The plants therefore get deformed on account of growth on the leeward side only.

High winds cause perhaps greatest injury to plants growing at high altitude on the Himalayas. At a low temperature obtainable in those regions, water freezes and its absorption becomes extremely low. On the top of it transpiration is very much increased by the wind. Under severe desiccation the growing buds die and thus the surviving plants are dwarfened. Trees and higher plants cannot grow at higher altitudes. This limit or the tree line is reached at about 12-14 thousand feet on the western Himalayas and about 2-3 thousand feet higher on the eastern Himalayas.

Anemophily, or pollination by wind and dispersal of seeds are beneficial effects of wind on vegetation. Most of the deciduous trees in tropical and subtropical India flower and fruit upon defoliation when the velocity of wind is the maximum in the summer.

18

The Biotic Factor

The plants and animals living together influence each other's life and also modify the environmental factors. In fact living organisms are most potent factors in the growth and development of each other. Pollination, seed dispersal, grazing, browsing, parasitism, symbiosis, etc. are some examples of such interactions.

The effect of man upon vegetation has, however, always been a dominating factor under the biotic influences. Indeed present vegetation of India is not what it would have been if the human interference had been less intense or absent. India has a very ancient history and throughout the centuries, she has been invaded in succession through the well known mountain passes along the north-west mountain chains of the country, and from the sea. These invasions had disastrous effects on the vegetation of the country and invariably natural forests were heavily felled, cleared and burnt, thus disturbing an equilibrium of primeval people and forests.

The earliest invaders who came to this country nearly 4 thousand years ago, were Aryans. They are believed to have been agriculturists and pastoral people; therefore, they destroyed and burnt a large part of the country's forests. The earliest known settlement of the Aryans were along the Indus from where they gradually spread towards the interior of the Punjab and formed their first empire, between the rivers Ghaggar and Saraswati, which is now represented by the districts of Ambala and Karnal. The Aryans advanced later into the plains of the Ganges and the Jamuna, spreading as far south as the Vindhyas. They had no less than 12 kingdoms and extended further south up to the rivers Mahanadi and Godavari, and eastwards to Bihar.

Stebbing (1921) has given an interesting history of the forest vegetation of India and according to him the great Khandava forest between the Ganges and the Jamuna rivers was burned during the Mahabharta by early settlers of the country. It is stated that the forest was burnt with great difficulty as frequent and abundant rain was again and again putting out the fire. Large areas were cleared for cultivation and by the time of Ramayana there had occurred in the country severe droughts and famine conditions.

The history of the early Aryan civilization is revealed in numerous archeological finds in Sind and Punjab, of which Harappa and Mohenjadaro are two of the well-known sites. The well built cities and the amount of burnt bricks used in their construction seem to show that the desert area in which there is now only thorny scrub was more hospitable to plant life in the past. Chowdhury and Ghosh (1921) after studying some plant remains from Harappa have shown that even by Harappa times the vegetation had already degraded to a scrub forest in which pockets of marsh land and tall grasses were present.

Some evidence of the luxuriance of the past vegetation may be seen in the ruins of the oldest city of Delhi, which was built by Pandvas nearly 15 centuries B.C. The site of this city is at the Porana Qila which was built later by Shershah.

In 330 B.C., Alexander the great invaded India and at this time the vegetation near about the Jehlum and the Salt Range country consisted of thick forests which had shielded the movements of Alexander's army. The destruction of forest vegetation went apace during the period succeeding Alexander, when in 250 B.C. Mauryan Empire was formed. This empire extended into Afghanistan and is famous by the names of Chandra Gupta and Ashoka the great.

Then there followed the Sunger Empire in 184 B.C.; the

Kushan Empire in 40 A. D. to 162 A.D.; the Gupta Empire between 320 A. D. and 490 A. D.; the Harsha Empire in 640 A. D.; which extended the Indian Empire to much beyond the limits of India on the north west to Afghanistan and Baluchistan, but the vegetation of the country was repeatedly destroyed and burnt and suffered a great set back.

After 850 A. D., India witnessed the rise of the Rajputs and then there occurred a procession of Muslim invasions from the Ghaznavis, the Ghoris and other slave kings (as Khiljis, Tughlaqs, etc.). India was then connected with the Mediterranean countries—Arabia and Egypt. Such human invasions perhaps destroyed the indigenous vegetation of the north-west India and opened the country for the invasion of Mediterrenean species brought to this country. The vegetation of north-west India and Pakistan in Sindh, and Rajputana has a number of Arabian and African species, whose spread and establishment were probably facilitated by clearing of the indigenous vegetation and reduction of plant competition.

By about the 11th century A.D., the Mohammedan invaders brought further destruction to the vegetation of North-west India. These people from Central Asia brought with them herds of cattle and clearing and burning of forests for the creation of pastures went apace during their region. The native population was time and again pushed away by the invaders into forests and in order to make their bread quickly they resorted to burning, getting yield for a year or two, and then running to other areas followed by the invaders. The method of shifting cultivation which has persisted in some parts of the country even to this day, bringing large scale destruction of the natural vegetation, has, probably originated under these conditions.

The other effect of the invaders from the North-west was the development of land and sea trades between India and the Middle-East countries. At this time Malabar teak and sandal wood,rice, ivory and other products of the forests were exported to Babylon, Egypt, Greece and Rome and the trade with Central Asian countries as well as with China through Kashmir, Tibet and Afghanistan developed rapidly.

In 1398 A.D., Timur invaded India, followed by Babar in 1526 A. D. With the spread of the Moghal Empire the wars of the Moghals with Sivaji in 1680 A.D. and those with Sikhs in 1738 A. D. —1748 A. D. Indian vegetation suffered serious consequences.

In 1757, the British had fairly stabilised in South India and the Anglo-Maharatta wars in 1775 A.D. to 1818 A.D., Anglo-French wars between 1774 A.D. to 1761 A.D. and wars of Hyder Ali with Indian kings and the British between 1766 A.D. to 1799 A.D., made equally disastrous effects on the vegetation of Southern India and the Deccan.

By about 1823, the British had become supreme upto the Sutlej, in the N.W. and Brahmputra in the east, and the Anglo-Sikh wars between 1780 A.D. to 1850 A.D., the Indian Mutiny in 1857, and the Anglo-Afghan wars during the closing years of the last century had made further serious repercussions on the whole of Indian vegetation. Although authentic records of specific destruction of forests are lacking yet it can be surmised that they gradually dwindled from the west to the east during the last 4,000 years. As a result the climate of the denuded country has probably deteriorated. At least the change is seen in the westward drifting of most of the Indian rivers in the extra peninsular territory in the silting of river beds, the spread of aridity and desert conditions from the N.W. of India towards the Ganges valley.

Shifting cultivation—the clearing of vegetation and then abandoning the areas—is clearly seen in the forests of Northern India, and the small patches of relict vegetation seen here and there, near old habitation sites, or associated with religious areas (Puri, 1954), show that the present vegetation of the country is certainly not what it was but has been shaped excessively by the adverse waves of human influences through the centuries. The advancing civilisation and culture of the country made heavy demands upon the forest wealth and there was ruthless exploitation without proper management.

The destruction and ruin of forest vegetation did not decrease even with the coming in of the British until some stability was restored and laws for the preservation of the forest wealth were enacted. The position of the Indian vegetation by about the end of the 18th century is summed up by Stebbing in the following words; ' Certain trees, the timber of which had a recognised valuc as articles for a valuable export timber trade, a trade which had existed in some cases for many centuries, had been proclaimed "royal trees" by the rulers of the territories in which they flourished. With these few excceptions the forests were free to all, although nominally they belonged to the ruler of the territory. This later claim existed, with certain small exceptions of forest territory, throughout the country as a whole, and formed the basis upon which it was ultimately found possible to build up the forestry department under the British regime."

"The forest area had been immensely reduced by fire and by the almost universal practice of shifting cultivation, this reduction having reached a point at which it was seriously jeopardising in parts and pursuits and business of the population, which was in the main agricultural and pastoral one. For the unchecked destruction of the forests was having an adverse effect on the maintenance of the water supplies over considerable areas in some parts."

"But the British of the period were unacquainted with the principles of forestry science. The report on the forests in the neighbourhood of the part of the country we first occupied were unanimous in regarding them as inexhaustible, so far as the materials that were required from them was concerned. The first decades of British occupation of India, therfore, witnessed no check, but rather an enhanced rate in the destruction of fine timber forests in these regions. And the destruction of the forests throughout the remainder, the great bulk, of the country, continued on the lines which had been in force throughout the centuries".

In the Himalayas extensive forests of deodar and in south of teak were ruthlessly exploited by the East India Company. The history of India has thus been one of excessive exploitation of Indian vegetation. Except for some forests at very high altitudes in the Himalayas and thick impenetrable forests in the Deccan, the great majority of Indian vegetation has been modified on account of the activities of man, which have been responsible in checking the natural line of succession. Much of the Indian forests are, therefore, not climatic climax but represent only bio-edaphic communities, sub-climax or post-climax, or relicts which are maintained in their present state by the biotic factors in operation since the centuries.

The influences of man upon vegetation may be classified under four main heads :—

1. Direct and indirect effects of wild and domesticated animals.
2. Excessive felling.
3. Burning.
4. Planting of indigenous and exotic species.

Direct and indirect effects of wild and domesticated animals

A great deal of damage to forests of India is done by grazing animals, most of which are domesticated; although a number of them are either let loose in the forest, or have strayed from domestication. It is estimated that of the total cattle population of India as much as 5% graze in forest areas. The 1935 census has revealed their number to be 295,280,000 heads of cattle. The grazing is more intense near villages, and invariably the fringes of forests are overgrazed. The grazing animals are not limited to the plains but on the Himalayas they are taken to alpine meadows above an altitude of 8,000-9,000 ft. during summer months. They sometimes play havoc to adjoining forest areas and during their movement do a great deal of damage to the entire forest belt. The destruction done by migratory herds of sheep and goats in innner dry valleys, e.g. Sutlej, are tremendous. Here every kind of plant, even *Artemisia maritima* is eaten up and there is hardly any ground flora left in *Quercus ilex* or *Pinus gerardiana* forests.

The common grazing animals in this country are cows, buffaloes, horses and donkeys.

The most injurious effect of grazing is on tree seedling

growth. The natural vegetation is changed from forest to scrub and then to a treeless grassy area if it is allowed free access to cattle. By trampling, the soil becomes compact and hardened and thus rendered unsuitable for the growth of trees. At some places grassy patches are filled with worthless thorny bushes and coarse weeds which are not eaten by the animals and thrive at the expense of fine grass and tree seedlings of useful species. In addition to the direct evil effects of grazing animals on vegetation lopping is done to provide the animals with other types of feed. Oak is the commonest species in the Himalayas which is lopped for this purpose and this has resulted in an almost disappearance of the species from some of important forest areas. In the forests of Indo-Gangetic plains lopping of species of *Albizzia, Bauhinia, Terminalia. Dalbergia, Ficus, Ougeinia* etc. is allowed to *gujars*.

In addition to grazing, goats, sheep and camels browse on shoots and leaves of higher vegetation. Goats and sheep are the worst enemies of forest vegetation and as a result of the excessive browsing large areas of the country have become denuded of vegetation and susceptible to erosion. Camels browse on almost everything in the Rajasthan, States of Saurashtra, U. P., Delhi. Punjab; and desert conditions, in these regions are partly accentuated by the ill-effects of excessive browsing by this animal. The population of these animals has been going up with the population of man and since he has been killing mostly wild carnivorous animals the population of the wild herbivorous animals has also been rising. It is only with the decline of vegetation that the entire animal population falls.

In addition to the domesticated animals, forests suffer from browsing of herbivorous wild animals *Sambur, Nilgai* and other members of the deer family browse on a number of tree species damaging them considerably. *Morus laevigata* and *Artocarpus chaplasha* are heavily browsed by deer in Assam and Bengal. These animals also perfer leaves and tender shoots of *Chloroxylon swietenia,* sandal, *Dalbergia latifolia, Pterocarpus marsupium, Albizzia lebbek* and *Hopea parvifolia.*

Shorea robusta seedlings are heavily damaged by *Nilgai* in Gorakhpur. Protection against these animals is rather difficult. Numerous types of hedges have been tried with partial success and the idea of live hedges has become popular in Madras. The plants which have been found more suitable in that state for live hedges are *Acacia planiforns*, *Euphorbia* spp., *Prosopis juliflora* and *Protium caudatum*. In forest plantations *Cassia siamea* was found to be successful in some cases in protecting the species inside its fence from browsing animals.

In addition to the above, bisons and elephants cause considerable damage. Seedlings of *Artocarpus hirsuta*, *Dalbergia latifolia*, *Pterocarpus marsupium*, etc. are repeatedly browsed by bisons in evergreen forests. Elephants feed on *Morus laevigata*, *Ficus cunia*, *Ficus glomerata*, other *Ficus* spp., *Artocarpus chaplasha* and bamboos. Bamboos are the favourite food of elephants and when growing in teak forests elephants destroy a good deal of teak. Elephants strip the bark of *Artocarpus integrifolia* in Malabar and destroy teak in Coimbatore and Malabar districts.

Pigs are another source of damage to forest species and uproot seedlings and damage coppice shoots. Teak stumps are usually uprooted to great extent by pigs.

In conifer forests, bark of *Cedrus deodara*, spruce and blue pine is stripped off by black bear who licks the rising sap and cambium layer beneath the bark. In this way a number of trees in these forests are girdled.

Seeds and seedlings of a number of species are eaten by hares, for example, sandal seedlings in Madras are greatly damaged. Porcupine usually gnaw roots of *Bombax malabaricum* and numerous other species in forests of Dehra Dun and Saharanpur. Rats are another pest in some forest areas. The last but not the least type of animals causing damage to Indian vegetation are the monkeys. They eat buds and young shoots of a number of species and also pull out seedlings. Young buds of *Artocarpus chaplasha* are the favourite food of monkeys.

The natural forest environment equilibrium is greatly disturbed by the destruction of the forests and it has invariably resulted in arresting of succession along natural lines. Due to the effect of the animal factor secondary equilibria of lower and seral stages are set up. For example many grasslands in this country are controlled by biotic influences of which grazing is one of the chief factors. Pandeya (1954) has shown in Madhya Pradesh that moderate grazing is necessary to keep out tree species from grasslands and maintain a healthy growth of palatable grasses like *Dichanthium* spp., *Bothriochloa pertusa*, *Iseilema anthephoroides*, etc. Intense grazing leads to replacement of these species by rough grasses and weeds. Further, the grasses show different growth forms according to the intensity of grazing.

The creation of desert conditions and spread of aridity in some parts of northern India are due partly to the biotic factor of which unrestricted grazing is the chief agent.

Grazed and ungrazed areas after a time develop into different biological systems and not only show differences in frequency of plants but also differences in soil conditions. For example, moisture content of the soil in grazed area becomes low. The amount of litter, organic matter and nitrogen decreases in heavily grazed soils and as a result of such changes only a specialised type of vegetation can grow.

After a ten year study of grazed and ungrazed plots in north-eastern Ohio, Dambach (1944) found that in ungrazed plots the number of seedlings, their growth and vitality was invariably better than in grazed plots. The soil factors in grazed and ungrazed plots is given in table 111.

TABLE III
Soil factors of grazed and ungrazed tracts
(After Dambatch, 1944).

	Ungrazed tract.		Grazed tract.	
	Surface cores.	6 in. cores.	Surface cores.	6 in. cores.
Average percolation rate cubic inches per hour.	31.4	0.08	2.59	0.11
Average volume weight of soil gms. per c.c.	1.12	1.58	1.07	1.46
Average percent by volume non-capillary porosity.	8.89	3.41	6.86	5.08
Average percent by volume capillary porosity.	41.65	37.08	43.65	36.33
Average percent total porosity or water holding capacity.	50.54	40.49	50.51	41.87

Although, grazing has in many cases resulted in forest destruction, it has sometimes bestowed beneficial effects. Grazing has been advocated as good in inducing regeneration of silver fir in some forests of Bashahr. Here sheep and goat eat away evergreen weeds like *Strobilanthes*, *Impatiens*, etc. disturb the thick layers of humus creating conditions suitable for regeneration of silver fir.

2. **Excessive felling.**

Before the scientific management started felling of trees in forest areas had been indiscriminate for most of the time. Selective felling of useful species has also been wholesale without providing for their regeneration. It is only under modern management systems that the growth of the species felled has been assured. The tendency of building up larger towns from prehistoric day must have necessarily led to the felling of timber trees and the present pattern of their distribution in the country might be traced accordingly.

Felling of climax vegetation has resulted in the initiation of secondary and deflected succession. Numerous examples can be given to show that the invasion of conifers in the Himalayas has been due to felling of climax oak forests. The felling in *Quercus semecarpifolia* forests on scarp slope in the Bashahar

Himalayas has brought in the community of *Pinus excelsa* and *Picea smithiana*. On the dip slope it has brought in *Abies webbiana* and numerous broad-leaved specios of *Populus cliata, Prunus padus, Salix* spp. *Acer* spp. etc.

The felling in *Quercus ilex* forests in inner dry valleys has brought in *Pinus gerardiana* and *Cedrus deodara*. The occurrence of shrubs of species of *Prusepia, Berberis, Morus, Pyrus* etc. in the Himalayan forests can be directly traced to indiscriminate felling.

In South India clear-felling of forests has enabled the spread of *Lantana camara* and in sal forests of Assam and Bengal it has facilitated the spread of *Eupatorium odoratum* westwards.

Excessive felling of forests has a serious effect on the development of soils. It is believed that the creation of *Usar, reh* and *kallar* soils in some parts of the Indo-Gangetic plains and of desert conditions in the north west has been the result of exceessive fellings of indigenous vegetation.

3. **Burning.**

Attention has already been drawn to the process of shiftting cultivation which existed for a long time over most of the regions of the country. In order to clear a forest perhaps the easiest method found was to burn it wholesale especially when not much use for the timber was known to earlier men. 'Jhumming' in which the crop is raised upon the ash of the previous forest stand is still practised by backward jungle tribes in Assam and eastern parts of Madhya Pradesh.

Devastating forest fires very often arise without man's intention. Depending upon their severity the growth of plants and vegetation are arrested at various levels. Sometimes the soil may be so desiccated and burnt that no growth may come up for a number of years.

Rough grazing grounds are burnt sometimes for producing tender grasses. As the inflammable amount of litter in such cases is small the underground root stocks of grasses are saved and sprout up following monsoon showers.

Burning of slash is a usual practice in forestry. This probably has a beneficial effect on activation of nitrifying organisms; sometimes the growth in burnt plots has been found to be better than in the unburnt plots.

In chirpine forests burning is supposed to facilitate germination of seed. A light burn is also advocated to be of some use in a number of forests in the country.

4. Planting of indigenous and exotic species.

Man has necessarily thought of rehabilitating denuded areas with useful plants. According to the needs he has planted both indigenous and exotic species for fruits, shade, avenues, flowers, timber, fuel and other useful resources at different sites. Most of such plantations have been achieved on the basis of trial and error. Some of them have stood the test of time and are flourishing, others have succumbed at one stage or another due to modified conditions of the soil and microclimate. Large scale plantations of teak are available in Madras, M.P., Bombay, Bengal and eastern U.P.; *Casuarina* in Madras, *Eucalyptus* in South India, *Cryptomeria japonica* in Darjeeling Himalayas etc. *Gmelina arborea* plantations in Assam failed on account of insect attack in that region. Teak defoliator and *Loranthus* have been the cause of ruin of a number of teak plantations. *Casuarina equisetifolia* has also suffered heavily from *Trichosporium vesiculosum*. Merits of mixed versus pure plantations of teak have been discussed by a number of foresters and recently Puri (1954) has given an ecological method of the introduction of exotics in the country. Besides the climate, he has examined the effect of these plantations on soil conditions.

In Kashmir Himalayas *Acacia pseudorobinia* and Lambardy poplar have become more or less naturalised. Bamboo plantations are raised for paper and now *Broussonetia papyrifera* is finding much use for the same.

Symbiosis

The influence of man and herbivorous animals both domesticated and wild has no doubt been supreme in destroying the natural vegetation. However, the association of organisms in any community or biome is a potent biotic factor for its existence. For example, a large variety of insects are necessary for bringing about pollination in different species of plants; animals disperse their fruits and seed and in many species association of micorrhizal fungi is necessary for absorption of nutrients. The concentration of carbon dioxide and oxygen in the atmosphere so necessary for plant life, are largely influenced by the proximity of other organisms. Likewise the amount of heat, light, available nutrients, water, etc. all are influenced by other plants. Disease causing organisms take a heavy toll of plant life. Thus no individual whether plant or animal seems to be independent in a community. Clements and Shelford have rightly emphasized the importance of biologic environment which has the potentiality of influencing the physical environment of plants.

McDougall has taken into account all such biologic inter-relationships which according to him can be called symbiosis (Gr. *syn.* together,—*bios*—life). The following forms of symbiosis have been defined by him; examples taken are those of Daubenmire.

I—Disjunctive Symbiosis–(associated organisms not in constant contact).

A. **Social**—(no direct nutritional relationships). Effects of one plant upon another with respect to shade, air, moisture, etc. also the substrate relations of saprophytic organisms.

B. **Nutritive.**

1. *Antagonistic*: Herbivorous animals and their food plants; carnivorous plants and their prey.
2. *Reciprocal:* (A condition of reciprocal parasitism in which the advantages of the relationship exceed the disadvantages for both organisms).

    Animals effecting pollination and dissemination incidental to their activities in obtaining food from plants; fungi cultured and disseminated by insects that use them for food; agricultural pursuits of man; ants inhabiting hollow thorns or other organs of vascular plants which they defend against the attacks of other insects.

II—Conjunctive Symbiosis—(Dissimilar organisms living in contact with each other).

A. **Social**—Lianes and epiphytes using other plants for support; algae inhabiting the hollow interiors of *Azolla* leaves, the roots of Cycads, thallus of *Anthocros*, etc.

B. **Nutritive.**

1. *Antagonistic*

   Parasitism of plants by bacteria, fungi, protozoa, nematodes, insects, *cuscuta*, *Loranthus*, etc.

2. *Reciprocal*

   Lichens; micorrhizae; nitrogen fixing bacteria in roots, stems and leaves; algae inhabiting protozoa, coelenterates, molluscs, flat worms, etc.

Examples for each of the types of symbiosis given above can be multiplied. But the more important aspect of such associations is the struggle for existence and how it leads to elimination of one species by another in the community. In the tropical vegetation of India the role of ants, termites and insects is spectacular. Climatic conditions and seasonal drifts including changes in humidity and temperature lead to sudden increase in their populations. They feed upon the roots, leaves and fruits and the ants even store grass seeds in their nests for eating them later.

The invasion of *Lantana* in the forests and of several other unwanted species is a perpetual menace to the forests. A large number of exotic weeds like *Opuntia, Croton sparsiflorus, Argemone mexicana, Gomphrena decumbens*, etc. have invaded open areas in recent years. Water hyacinth has often threatened rice fields in parts of Assan, Bengal and Bihar.

The insects and plant pests have their own parasites and in recent years their biological control has been achieved. Thus introduction of the insect parasite on the *Opuntia* from Mexico to Australia and India has controlled growth of the host weed.

BIBLIOGRAPHY

Allee, W. C. 1926. Measurement of environmental factors in the tropical rain forest of Panama. *Ecology* **7**: 273-302.

Auden, J. B. 1940. *Ecological lectures for Military Engineer Officers*, Delhi.

Bakshi, T. S. 1952. The autecology of *Anisochilus eriocephalus* Benth. *J. Indian bot. Soc.* **31**: 269-280.

Banerji, J. 1952. Discussion on the role of vegetation in desert control. *Ind. For.* **78** : 376-378.

Beaufort, A. 1805. *See in* Finch & Trewartha 1942, pp. 67-68.

Bharucha, F. R. and D. B. Ferreira, 1941. The biological spectra of the Matheran and Mahabaleshwar Flora. *J. Indian bot. Soc.* **20**: 195-211.

Bhatia, K. K. 1954. Ecology of teak, *Tectona grandis*. *Ph.D. thesis*. Saugar Univ.

Blanford, H. B. 1889. *Climates and weathers of India, Ceylon and Burma and the storms of Indian seas*, London.

Bor, N. L. 1938. A sketch of the vegetation of the Aka hills, a synecological study. *Ind. For. Rec.* (Bot.) **1**: 103-221.

——— 1938. The vegetation of the Nilgiris. *Ind. For.* **64**: 600-609.

——— 1939. Some remarks upon the geology and flora of the Naga and Khasi hills. *Roy. Bot. Garden, Calcutta., 150th Anniv. vol* : 129-137.

——— 1942. Ecology, Theory and Practice. *Proc. Ind. Sci. Cong. Presid. add.* (Botany).

——— 1942. Relict Vegetation of the Shillong plateau, Assam. *Ind. For. Rec.* (Bot.) **3**: 152-195.

——— 1947. Glossary of terms used in plant ecology. *Ind. For. Rec.* (N.S.) **6** :

Bournebusch, C. H. 1930. *The fauna of forest soil*. Copenhagen.

Burkholder, P. R. 1936. The role of light in the life ot plants. *Bot. Rev.* **2** : 1-52, 97-172.

Burns, W., L.B. Kulkarni, & S.R. Godbole, 1931. Succession in Xerophytic Indian grassland. *J. Eco.* **19** : 389-391.

Busgen, M. & E. Munch, 1929. *Structure and life of forest trees*. London.

Cain, S. A. 1947. Characteristics of natural areas and factors in their devolopment. *Ecol. Monog.* **17** : 185-200.

Calder, C. C. 1937. An outline of the vegetation of India. In an outline of the field sciences of India. *Proc. 25th Ind. Sci. Cong*: (*Abstract*) 71-91.

Chakradeo, G. M. 1922. An ecological study of Deccan grassland II. *Proc. 9th Ind. Sci. Cong:* (*Abstract*): 119-120.

——— 1926. Methods of grassland improvement (A Study in Plant Sociology). *Proc. 13th Ind. Sci. Cong.* (*Abstract*): 227.

Champion, H. G. & I. D. Mahendru, 1933. Multiple yield table for deodar (*Cedrus deodara* Laws.) *Ind. For. Rec.* (O. S.), 8:

Champion, H. G. 1936. A preliminary study of the Forest types of India and Burma. *Ind. For. Rec.* (N. S.) 1: 1-286.

——— 1938. Progress of forestry in India during the past twenty five years: in *Progress of science in India during the past twenty five years. Proc. 25th Ind. Sci. Cong.* (*Abstract*): 434-456.

Chatterjee, D. 1939. Studies on the endemic flora of India and Burma. *J. Royal Asiatic Soc. Bengal,* 5: 19-67.

Chhiber, H. L. 1945. *India part I. Physical basis of geography of India.* Banaras.

Chourey, S. 1953. A study of the reproductive capacity of six dominant forest trees. *M. Sc, Thesis, Saugor Univ.* 1-90.

Chowdhury, K. A. & S. S. Ghosh, 1951. Plant remains from Harappa. *Ancient India.* 7 (1946).

Clark, O. R. 1940. Interception of rainfall by prairie grasses, weeds and certain crop plants. *Ecol. Monog.* 10: 243-277.

Clements, F. E. 1916. *Plant succession, an analysis of the development of vegetation.* Carnegie Institute of Washington. 512 pp.

Coile, T. S. 1940. Soil changes associated with loblolly pine succession on abandoned agricultural land of the Piedmont Plateau. *Duke Univ. School For. Bul.* 5. 85 pp.

Cooke, T. 1903–1908. *The Flora of the Presidency of Bombay. Vols. I & II.*

Cooper, W. S. 1917. Redwoods, rainfall and fog. *Plant World* 20: 179-189.

Coulter, J. M., *et al.* 1931. *Text book of Botany.* Vol. 3, *Ecology.* New York. 499 pp.

Coulson, A. L. 1940. *The geology of water supply in "Geological lectures for Military Officers."* Delhi.

Dansreau, P. 1951. Description and recording of vegetation upon a structural basis. *Ecology*. **32**: 172-229.

Daubenmire, R. F. 1950. *Plants and Environment*. New York. 424 pp.

Dudgeon, W. 1920. A contribution to the ecology of the Upper Gangatic Plain. *J. Indian bot. Soc.* **1**: 296-324.

Eliot, T. 1906. *Climatological atlas of India*. Edinburgh.

Evans, G. C. 1939. Ecological studies on the rain forest of Southern Nigeria II. The atmospheric environmental conditions. *J. Ecol.* **27**: 436-482.

Finch, C. F. & G. T. Trewartha, 1942. *Physical elements of Geography*. McGraw H Book Co., London.

Flahault, C. & C. Schroter, 1910. Phytogeographische Nomenklatur. Berichte and vorschlage. Proc. *III Intern. Bot. Congr.* Bruxelles.

Fox, C. S. 1936. Buchanan's laterite of Malabar and Kanara. *Rec. Geol. Surv. Ind.* **69**: 389-422.

Gams, H. 1918. Pruzipienfragen der vegetationsforschung. *Vierteli. Naturf. Ges.* Zurich. **63**,

Gleason, H. A. 1918. On the development of two plant associations of Northern Michigan. *Plant World*. **21**: 151-158.

——— 1927. Further views on the succession concept. *Ecology* **8**: 299 326.

——— 1929. Raunkiaers law of frequency, *Ecology* **10**: 406-408.

Gorrie, R. M. 1933. The Sutlej deodar, its ecology and timber production. *Ind. For. Rec.* (Bot.) **17**: 1-140.

Grasovsky, A. 1929. Some aspects of light in the forest. *Yale Univ. School For. Bul.* **23**: 2-53.

Griffith, A. L. & R. S. Gupta, 1947. Soils in relation to teak with special reference to laterisation. *Ind. For. Bul.* **14**.

——— 1947. Determination of the characteristics of soil suitable for sal (*Shorea robusta*). *Ind. For. Bul.* 138.

——— 1947. Recording of soil and site characteristics in the field. *Ind. For. Bul.* 135.

Grisebach, A. 1872. *Die vegetation der erde nach ihrer klimatischen Anordnung*. Leipzig.

Hann, J. Uon & V. Conard, 1936. *See in* Haurwitz & Austin 1944.

Haurwitz, B. & J. M. Austin, 1944. *Climatology*. McGraw-Hill Book Co. New York. pp. 23-27.

Hewetson, C.E. 1941. Observations on the ecology of *Tectona grandis* in the Central Provinces. *Ind. For.* **67**: 617-29.

——— 1953. A discussion on the ecological position of sal (*Shorea robusta*) in Central India. *Ind, For.* **79**: 310-320.

Hole, R. S. 1913. Note on the chief fodder grasses of Indian forests. *Ind. For.* **39**: 69-81.

Hooker, J. D. 1906. *A sketch of the flora of the British India.* Oxford.

Hoon, R. C. 1939. Study on the soils in the hill areas of Kashmir; an investigation of soil profiles under deodar, blue pine, silver fir and chir. *Ind. For. Rec.* (Bot.) **3**: 195-262.

Hosking, J. S. 1937. The ratio of precipitation to saturation deficiency of the atmosphere in India. *Curr. Sci.* **5**:

Hurst. F. B. 1951. Climates prevailing in the yellow gray earth and yellow brown earth zones in New Zealand. *Soil Sci.* **72**: 1-19.

Joachim, A.W.R. 1937. Studies on Ceylon soils. *Trop. Agric.* **88**: 12.

Kadambi, K. 1943. Valuation survey by spot method. *Ind. For.* **69**: 107-112.

Kendrew, W. G. 1930. *Climate.* Oxford.

——— 1938. *Climates of the Continent.* Oxford.

——— 1949. *Climatology treated mainly in relation to distribution in time and space.* Oxford.

Kimball, G.W. & E.E. Carter, 1913. Influence of shade and other factors on plantations. *For. Quart.* **11**: 176-186.

Klages. K.H.W. 1942. *Ecological crop geography.* Mac-Millan New York. 615 pp.

Koppen, W. 1900. Versuch einer klassifications der klimate, vorzugsweise nach iheren Beziehungen zur Pflanzenwelt. *Geogr. Zeitscher.* **6**: 593-611 and 657-679.

——— 1923. *Die Klimate der Erde.* Berlin.

——— & Geiger, R. 1930. *Handbuch der klimatologie.* Gebruder Borntraeger, Berlin.

Kramer, P.J. 1949. *Plant and soil water relationships.* Mc-Graw Hill Co., New York. 345 pp.

Krishnan, M. S. 1952. *Geology of India and Burma.* Madras.

Krishnaswamy, V.S. & G. S. Puri. 1954. Results of an experiment to study the succession of ground flora species under forest plantation raised on old agricultural land in the New Forest, Dehra Dun, India. *Ind. For.* **80**: 522—530, 633-648.

Kuchler, A. W. 1951. The relation between classifying and mapping vegetation. *Ecology* **32**: 275-283.

Lang, R. 1915. *See in* Hosking, J.S., 1937.

——— 1926. *Forstlicha Standortslehra*, Berlin.

Larsen, J. A. 1940. Effect of site factor variations and responses in temporary forest type in northern India. *Ecol. Monog.* **10**: 1-54.

Laurie, M. V. & Sant Ram 1940. Yield and stand table for teak (*Tectona grandis*) plantations in India and Burma. *Ind. For. Rec.* (N. S.) **4A**: 1.

Love, A. & D. Love. 1943. The significance of differences in the distribution of diploids and polyploids. *Heredetas*, **29**:

——— 1948. Chromosome numbers of northern plant species. *University Institute of Applied Sciences Department of Agriculture Reports, Series B. No. 3.*

Lowdermilk. W. C. 1927. A method for rapid survey of vegetation. *J. For.* **25**: 181-185.

Lundegardh, H. 1931. *Environment and plant development*, London. 330 pp.

MacLean, R. C. 1919. Studies on the ecology of tropical rain forest. *J. Ecol.* **7**: 5-171.

Maximov, N. A. 1929. *The plant in relation to water*, London. 451 pp.

Meyer, A. 1926. Uber einige zusammenhange zwischen klima und Boden in Europa. *Cehm. Erd.* **2**: 209-347.

Misra, R. D. 1938. Edaphic factors in the distribution of aquatic plants in the English Lakes. J. Eco. **26**: 411-461.

Misra, R. 1944. Variation of leaf form in *Potamogeton perfoliatus* L. *J. Indian bot. Soc.* **23**: 44-52.

——— 1944. The vegetation of the Rajghat Ravines. *J. Indian bot. Soc.* **23**: 113-121.

——— 1945. A botanical chart of the Banaras Hindu University grounds. J. *Ben. Hin. Univ.* **9**: 49–72.

——— 1945. The soil complex as studied in plant ecology. *J. Ben. Hin. Univ.* 9: 13-16.

——— 1946. An ecological study of the vegetation of the Benaras Hindu University grounds. *J. Indian bot. Soc.* 25: 39-59.

——— 1946. A study in the ecology of lowlying lands. *Ind. Ecol.* 1: 27-46.

——— 1949. Basic concepts of ecology. *Bul. Bot. Soc. Saugor Univ.* 1: 13-14.

——— 1950. Modern approach to the study of ecology. *J. Biol Soc. Assam* 1: 8-9.

——— & N.K. Joshi. 1952. The forest complex of Patharia Hills. *J. Indian bot. Soc.* 31: 154-170.

——— & G. S. Puri. 1955. Progress of plant ecological studies in India, Burma and Ceylon. Unpublished.

——— & B. S. Siva Rao. 1948. A study in the autecology of *Lindenbergia polyantha* Royle. *J. Indian bot. Soc.* 27: 186-199.

Mohan, N. P. 1933. Ecology of *Pinus longifolia* with particular reference to Kangra and Hoshiarpur Forest divisions. *Proc. Punj. For. Conf.* Lahore.

——— & Puri, G. S. 1955. The Himalayan conifers III. The succession of forest communities in Oak-conifer forests of the Bashahar Himalayas. *Ind. For.* 81: 465-562; 646-711.

Mooney, H. F. 1933. The forest of the Orissa State. *Ind. For.* 59: 200-221.

——— 1938. A synecological study of the forests of Western Singhbhum with special reference to their geology. *Ind. For. Rec.* (N. S). Botany 2:

——— 1942. A sketch of the flora of the Bailadela range in Bastar State. *Ind. For. Rec.* (N. S). Botany 3, 7.

——— 1947. The occurrence of some indigenous species of Rosaceae in Bihar, Orissa and the neighbouring States *J. Indian bot. Soc.* 26: 75-82.

Morrison, C.G.T., A. C. Hoyle, & J. F. Hope. 1948. Tropical soil vegetation catenas and mosaics. A study in South Western part of the Anglo-Egyptian Sudan. *Eco.* 36: 1-84.

Navalkar, B. S. 1940. Studies in the Ecology of Mangroves. I. *J. Univ. Bomb.* 7: 5.

——— 1941. Studies in the Ecology of Mangroves. II *J. Univ. Bomb.* 7: 5.

——— 1941. Studies in the Ecology of Mangroves. II. *J. Univ. Bomb.* 9: 5.

——— & F. R. Bharucha. 1948. Studies in the Ecology of Mangroves. IV *J. Univ. Bomb. N.S.* **16**: 35-45.

——— 1949. Studies in the Ecology of Mangroves V. *J. Univ. Bomb.* (N. S.) **18**: 17-35.

——— 1949. Studies in the Ecology of Mangroves. VI. *J. Univ. Bomb.* **18**: 7-16.

Normand, C.W.B. 1937. The Weathers of India. (*In an outline of the field Sciences of India.*) *Ind. Sci. Cong. Assoc.* 1-16 pp.

Oosting, H. J, 1938. *The study of plant communities.* San Fransisco, California, 389 pp.

Pandeya, S. C. 1954. Grassland ecology. *Ph. D. Thesis* Saugar Univ.

Park, O. 1931. Studies in the ecology of Coleoptera II. The relation of certain coleoptera to plants for food and shelters specially those species associated with fungi in Chicago area. *Eco.* **12**: 188-207.

Paulsen, O. 1915. Some remarks on the desert vegetation of America. *The Plant World* **18**: 155-161.

Pearsall, W. H. 1920. The aquatic vegetation of the English Lakes. *J. Eco.* **8**: 163-199.

——— 1938. Soil complex in relation to plant communities *J. Eco.* **26**: 180-193; 194-201; 299-315.

——— 1945. Leaf fall in Hartfordshire woodlands. *Trans. Hartford, Nat. Hist. Soc.* **22**: 98-97.

——— 1954. *Mountains and Moorlands.* London.

——— & C.H. Mortimer. 1939. Oxidation-reduction potentials in water logged soils, natural waters and mud. *Eco.* **27**: 483-501.

Pelton, J. 1953. Ecological life cycle of seed plants. *Eco.* **34**: 619-628.

Phadnis, S. M. 1926. A note on the succession of species leading to grasslands on a trap hill. *Proc. 13th Ind.. Sci. Cong. Abstract.* 215.

Prescott, J. A. 1949. A climatic index for the leaching factor in soil formation. *J. Soil Sci.* **1**:

Puri, G. S. 1948. The flora of Karewa series of Kashmir and its phytogeographical affinities with chapters on the methods used in identification. *Ind. For.* **74**: 105-122; 152-164; 210-225; 240-244.

——— 1950. Forest ecology and evaporation measurements in India. *Sci. & Culture* **15**: 465-467.

——— 1950. Soil pH and forest communities in the Sal (*Shorea robusta*) forests of Dehra Dun Valley, U. P., India. *Ind. For.* **76**: 292-609.

——— 1950. The distribution of conifers in the Kulu Himalayas with special relation to geology. *Ind. For.* **76**: 144-153.

——— 1951. Advances in the ecology of teak (*Tectona grandis*) in India. *Proc. VIII Silvicultural Conference*, Dehra Dun.

——— 1951. Forest ecological problems in India. *Proc. VIII Silvicultural Conference*, Dehra Dun.

——— 1951. The amount of foliar ash in Sal (*Shorea robusta*) trees of different quality classes in India. *J. Indian bot. Soc.* **31**: 82–86.

——— 1952-53. Problems in the ecology of sal (*Shorea robusta*) in Madhya Pradash. *Bul. Bot. Soc.* Saugor Univ. **5**: 24-31.

——— 1954. Land erosion—a problem in dynamic ecology. Bull 21, *Nat.Geog. Soc.* India, 6-21.

——— 1954. Soil climate in some Indian forests. *Indian bot. Soc.*, **23**: 394-416.

——— 1954. Plant ecological research in India. *Ind. For.* **8**: 435-441.

——— 1954. The ecological technique of the introduction of exotic trees in India. *J. Indian bot. Soc.* **33**: 8-18.

——— 1954. Forest-types studies in India. *J. Indian bot. Soc.* **33**: 19-27.

——— & A. C. Gupta. 1951. Foliar ash and Cao in sal and associated vegetation in the Dun Valley. *J. Indian bot. Soc.* **29**: 139-144.

——— 1951. The Himalayan conifers II : The ecology of humus in conifer forests of Kulu Himalayas. *Ind. For.* **77**: 55-63; 124-129.

——— & R. Misra 1954. Ecology of Indian trees. *Sci. & Culture* **20**: 249-250.

——— 1954. On botanical teaching and research. *Sci. & Culture* **20**: 431-433.

Ramdas, L. A. 1949. Evaporation measurements in India. *Central Board of Irrigation J.* **6**.

Ranganathan, C. R. 1934. Studies in the ecology of Shola grassland vegetation of the Nilgiri plateau. *Ind. For.* **64**: 523-541.

Raunkiaer, C. 1934. *The life form of plants and statistical plant geography*. Oxford.

Raychaudhry, S. P. 1937. On the nature of laterite and lateritic soils. *Sci. & Culture* **3**: 69-74.

—— 1937. Studies on Indian red soils II. Fixation of Phosphates. *Ind. J. Agric. Sci*, **11**: 205-242.

—— 1943. Physico-chemical and mineralogical studies of black and red soil profiles near Coimbatore. *Ind. J. Agric. Sci.* **13**: 264-272.

Rayner, M. C. & Jones W. Neilson, 1946. *Problems in tree nutrition*. London.

Rosayro, R. A. de 1943. The soils and ecology of the wet evergreen forests of Ceylon I, II. *Trop. Agri.*, **98**: 4-14.

—— 1951. The ecology of the patanas. *Ceylon Assoc. Sci. Section B.*

Rowe, J. S. 1947. Characterisation and comparison of forest stands by means of phytographs. *Silviculture Research Note 93.*

Rowntree, J. B., 1954. An introduction to the vegetation of the Assam Valley. *Ind. For. Rec.* (N. S.) **9**: 1-87.

Russel, E. J. 1950. *Soil conditions and plant growth*. Longmans.

Saksena, S. B. 1954. Ecological, morphological and taxonomical studies on soil microfungi of some forest soil types of Sagar. *Ph. D. Thesis*. Saugar University.

Salisbury, E. J. 1916 & 1918. The Oak-Horn bean woods of Hartfordshire. *J. Eco.* **4** : 83-117; **6**: 14-52.

—— 1921. The significance of the calcicolous habit. *J. Eco.* **8** : 202-215.

—— 1924. The influence of worms on soil reaction and the stratification in undisturbed soil. *J. Linn. Soc. Bot.* **46**:

—— 1929. The biological equipment of species in relation to competition. *J. Eco.* **17**: 197-222.

—— 1946. In symposium on "The ecology of closely allied species. *J. Eco.* **33**: 115-116.

Sarup, S. 1952. Plant ecology of Jodhpur and its neighbourhood. A contribution of the ecology of North Western Rajasthan. *Bull. Nat. Inst. Sci.*, No. 1. 223-232.

—— & S.P. Singh, 1953. An ecological survey of vegetation of Hastinapur. *Proc. 40th, Indian Sci. Congress, Abstract*: 105.

Saxton, W. T. 1922. Mixed formation in time: A new concept in ecology. *J. Indian bot. Soc.* 3: 30-33.

Schimper, A. F. W. 1903. *Plant geography.* Oxford.

——— & F. C. Von Faber, 1935. *Pflanzengeographic auf physiologicher Grund lage* (third edit.) Jena.

Sharma, V. B. 1953. Autecology of *Sida acuta* Burn, f. *M. Sc. Thesis.* Saugar University.

——— 1955. An autecological study of *Boswellia serrata* Roxb. *Ph. D. Thesis.* Saugar University

Shirely, H. L. 1929. The influence of light intensity and light quality upon the growth of plants. *Amer J. Bot.*, 16: 355-390.

——— 1929. Light requirements and its silvicultural practice. *J. Eco.* 27: 534-538.

Shrivastava, M.P. 1952. Autecological study of *Heylandia latebrosa* DC. *M.Sc. Thesis.* Saugar University.

Smith, W. G. 1913. Raunkiaer's "life forms" and statistical factors. *J. Eco.* 1: 16-26.

Stamp, L. D. 1925. The vegetation of Burma from an ecological stand point. *Univ. Rangoon Res. Mono.* 1.

Stebbing, E. P. 1921. *The forests of India.* 3 vols. London.

Stephens 1949. Taxonomic arrangements of soil. *J. Soil Sci.* 1.

Subrahmanyam, V.P. 1956. Climatic types of India according to the rational classification of Thornthwaite. *Indian J. Meteorology & Geophysics.* 7: 253-264.

Suri, P. N. 1933. A study in the ecology and silviculture of Himalayan *(Picea morinda)* and silver fir (*Abies pindrow*) with special reference to works in progress in Kulu. Parts I & II. *Proc. Punj. For. Cong.* Lahore. 51-88.

Tam, O. 1950. *Northern coniferous forest soils.* Oxford.

Taylor, W.P. et al. 1935. The relation of jack rabbits to grazing in southern Arizona. *J. For.* 33: 490-498.

Thornthwaite, C. W. 1931. The climate of North America according to a new classification. *Geog. Review,* 21: 633-655.

——— 1933. The climates of the earth. *Ibid.* 23: 433-470.

Tiwari, D.K. 1952. An ecological survey of natural grasslands with special reference to grass-legume association. *M. Sc. Thesis.* Saugar University.

——— 1955. A study of the bioecology of grasslands of Madhya Pradesh (India). *Ph. D. Thesis*. Saugar University.

Transeau, L. N. 1905. Forest centres of eastern America. *Amer. Nat.* 39: 875-889.

Trivedi, T.S. 1954. Notes on the autecology of *Sesbania bispinosa* Fawe. E. T. Rendle. *M. Sc. Thesis.* Saugar University.

Troup, R. S. 1921. *Silviculture of Indian trees.* Vol. I to III. Oxford.

Turesson, G. 1922. The genotypical response of the plant species to habitat. *Hereditas* 3: 211 350.

Wadia, D.N. 1935. *Geology of India.* Calcutta.

——— 1945. Soils of India. *J. Sci. Ind. Res.* 3: 359-367.

———, Krishnan, M. S. & Mukherjee 1932. Introductory note on the geological foundations of the soils of India. *Rec. Geol. Survey India,* **68:** 363-391.

Warming, E. 1895. *Plantesamfund, Grundtrak af den okologiska Plantegeografi.* Copenhagen.

——— 1909. *Ecology of plants.* Oxford.

Wherry, E. T. 1924. Soil acidity preferances of earthworms. *Ecology* 3: 346-347.

Zederbauer, C. 1908. The light requirements of forest trees and method of measuring light. *For. Quart* 6: 255-262.

Zon, R. 1911. New viewpoints in silviculture. *For. Quart.,* 9: 205-218.

——— 1913. Quality classes and forest types. *Soc. Amer. For. Proc.,* 1: 179-189.

A

B

M

N

O

P

Indian Manual of Plant Ecology

by R. Misra & G. S. Puri

ERRATA

Page	Line	Read	Instead of
i	15	field	fleld
iv	5 from last	obliged	all of obliged
2	8	Palaeo-ecology	Paleo-ecology
3	25	It	In
9	21	In	n
13	9	Reproductive	Reproduction
16	36	programme	programe
17	last	*oleracea*	*oleraceae*
18	2	*annuum*	*annum*
18	12	because	gecause
20	7	Pradesh. *Strobilanthes*	Pradesh *Strobilanthes*
21	2	the plant community	the community
22	1	help	helps
22	2	gets	get
22	13	trees, grassy	trees; grassy
22	15	vegetation, and	vegetation : and
23	7	topmost	top most
23	19	respect, some	respect, however, some
23	28	of plant	of a plant
25	10	procession	processions
26	8 from last	A-E seral	A. E seral
26	8 from last	climax community	climax communities
27	29	C/N ratio in	C/N in
28	13	rhythmic	rhythmical
28	33	attains	attain
30	30	withstand	withst and
32	31	dominant ; in	dominant, in
33	33	Himalayas	Himalayas,
36	14	Suffix 'ation'	Suffixation
36	14	Suffix 'ies'	Suffixes
37	last	into	in
38	11	Gleason	Gleeson
38	13	fluctuating	uctuating
38	19	'Synusium'	Synusium
41	19	On	In
42	5	towards one	towards. one
53	9 from last	features	feautures
55	vertical	Percent	Pereent
56	8	of	or
58	15	Ephemerals	Ephemeral
60	7 from last	The degree of constance is also expressed on the scale of presence as defined above provided unit area is studied in each of the stands.	If a species is present in 81 to 100% of the quadrats studied then it is said to be constantly present.
61	13	Exclusive	Exclusives
62	8 from last	carbon	Carbon
62	2 from last	practical	practicable

64	8 from last	into	into
66	3 from last	The	Tbe
67	3	(Ph.	Ph.
67	11	Raunkiaer	Raunkiaer's
68	3	forms :	forms,
72	6	them	these
72	6	these	them
76	25	it (Fig- 16).	it. (Fig. 16).
76	34	10·2·0	10-.10
76	35	2·00 0·20	2,00 2 20
76	36	020-0·02	020·0 00
78	2	rocks on	rockson
79			
1st col.	7 from last	ferous	ferious
81			
3rd col.	2 from last	form	from
11th col.	17	porous, soft,	in colours porous, soft
,, ,,	31	rough	trough
85	9 from last	amount	amounts
88	8	form	from
88	19	form	from
90	2 from last	impervious, liberate	impervious. liberate
91	9 from last	*Avicennia*	*Avicenia*
91	last	Examples	Exmples
93	14	gravelly, and are	gravelly, are
94	5	other	others
98	4 from last	Table 9. Jensen	Table 9 Jensen
104	1	13	12 (Concluded)
106	5	from	From
108	17	fungi, bacteria	fungi bacteria
115	4 & 5 from last	in his study of some	Study of some in his
119	4	Ca below	Cabelow
121	6	to	and
121	7 from last	*Chara*	*Chaera*
125	3	figure	figures
126	5 from last	Status	States
131	2 from last	cells, and provides	cells, provides
132	17	influences the interception	infiuence the intercepting
133	15	26 to 73%	26—73%
133	22	(Table 26)	(Table 26.)
133	5 from last	regions of.	regions; of
135	11 from last	*B. Dead Plants*	*B. Dead Plant*
135	6 from last	capacity	ca pacity
140	14	Fig. 23a	Fig. 23
140	18	of mesophytic type as	as mesophytic type of
140	3 from last	Fig. 23a	Fig. 23
152	2	desiccation	dessication
152	24	sulphide) are	sulphide, are
153	2	Soil redox	Soils red ox
153	5 & 6	Thiocynate	Thyiocynate
156	1	*Helophytes*	*Halophytes*
156	20	*Salvinia*	*Salvia*
157	19	algae	Alage
158	23	xerophytism	Xerophytism
163	24	(High rainfall).	High (rainfall)
163	last	Percentage. 0.0069 0.2834	Percentage. 0.0069
164	21	change change change	changing changing changing

164	11 from last	Nitrites	Nitrates
165	7	*Dalbergia*	*Da ergia*
165	13 from last	Hotel, Puri	Hotel Puri,
166	9–11	Salt content Parts per 100,000 Calcium | Sodium	Salt content Calcium | (Parts per 100,000 sodium
166	2 from last	analyses	analysis
169	21	10.51	12 51
170	4	near an industrial	near the industrial
171	15	Carbonated	Carborated
171	last	Basin where	Basin. where
172	6	complex,	complex
172	9 from last	removes	remove
174	4 from last	soil air and	soil and
175	3	adapted	adopted
178	6	are	were
178	19	correlation	co-relation
182	8	*elluvial*	*alluvial*
182	Column 1	Solum	Colum
188	3	podsolisation.	podsolisation,
195	8	composed	compound
198	last	Chandler	Chander
199	,, 2; 22	accumulation	accumulatlon
199	,, 6; 27	calcification	caleificatlon
199	,, 8; 20	grains especially	grains, especially
199	,, 8; 22	Excellent	excellent
200	,, 4; 4	sub-humid	ub-humid
200	,, 4; 12	Effective	Effective
206	21	climate, as modified by him from	climate, modified from
206	22	are given	(*loc. cit.*) are given
207	2	are although	are, although
207	2 and 3	above yet latitude	above latitude
210	10 from last	develop at middle	develop middle
211	12	fire	fires
211	7 from last	1 to 3	3 to 1
216	11	factor	fatcor
219	6 from last	*latitudes*	*latitues*
219	last	Klages	Klagges
220	18	inspite	in spite
221	1	36a	36
221	last	36a	36
224	2 from last	39	30
225	2	their	th eir
226	last	225	223
227	23	*Herbaceous*	*He baecous*
229	6	boundaries	bounaries
233	1	vegetation	vegefation
233	1	Soil	Soile
234	23	P/s. d.	P/s. d
234	24	podzols, red	podzols red
234	3 from last	deficit	defict
236		Thornthwaite's	Thornth waite's
237	25	pedacals	pedocals

239	last	rainy	rain
240	6	temperature- precipita-	temp-precipita
240	19	De Martonne	De Mar onne
244	last	different	diffrnt
248	3 from last	The sudden change which is no mere intensification of the previous light sea breeze deserves consideration.	Why the sudden change, which is no mere intensification of the previous light sea breeze,
249	2	equatorial	Equatorial
249	3 from last	of	af
253	21	bryophytic	Bryophytic
257	14	when closer	when a closer
257	18	An	At
258	8	are	is
258	14, 15	Koppen	Koppen
260	5 from last	Carter,	carter
261	2	Champion	Campion
261	2	forests; climatic	forests, climatic
261	3	formulae	formulas
263	2 from last	Fig. 55	Fig. 52
265	2 from last	Thornthwaite's	Thornwaite's
265	last	per humid	Per humid
267	last	Lundegardh as shown in	Lundegarh in
268	5	(in u)	in mm,
268	12	Circa 100	Circa 600
268	13	100	2 mm
269	12	loosing	losing
270	6	(in foot candles)	(in foot candles metre
271	10	of	af
271	12	Table 76.	Table 76
276	25	Number	Numbe
277	9	82. The	82 The
278	2	foot	food
279	2 from last	heating	hea ting
279	6 from last	value of insolation	value insolatien
280	20	mortality	morality
280	22	Shirley	Shirely
281	11	U.S.A. as in	U S A. in
282	10	forest	forests
283	3	lights	light
283	14	84	83
284	Column 4; 11	Maximum light in ft. candles	Maximum light in the candles
284	,, 6; 11	Minimum	Maximum
284	Col. 2; 15 from last	Colorado	Colorada
286	12	discussed on pages 17 and 18.	discussed than in those that grow in shade.
287	7 from last	and	ond
287	3 from last	temperature with	temperature and with
289	11	On	on
289	22	heat units	heat in
292	last	than	then
297	4	diameter in cm.	diametcr.
297	5	Altitude in m.	Altitude.
297	6 and 7	40	40 cm.

		41	41 cm.
297	9	38	38 cm.
299	13	according to rainfall	by altitude and site quality
310	3	temperature	tempeatre
310	6	more pronounced in	more in
314	20	plants serves for	plants is a
317	18	the	an
319	23	reign	region
321	9	exceptions	excceptions
321	11	latter	later
322	5 from last	*maritiana*	*maritima*
323	3 from last	prefer	per fer
323	last	*parviflora*	*porvifolia*
324	6	*Acacia planifrons*	*Acacia planiforus*
324	21	species; they	species and
325	last	are	is
326	22	and disturb	disturb
327	3	*ciliata*	*ciliata*
327	7 from last	is	are
329	4	mycorrhizal	micorrhizal
330	4	Lianas	Lianes
330	6	*Anthoceros*	*Anthocros*
330	12	mycorrhizae	micorrlizae